A Place like No Other

A Place like No Other

DISCOVERING THE SECRETS OF
SERENGETI

ANTHONY R. E. SINCLAIR

WITH RENÉ BEYERS

PRINCETON UNIVERSITY PRESS
PRINCETON & OXFORD

Published by Princeton University Press
41 William Street, Princeton, New Jersey 08540
6 Oxford Street, Woodstock, Oxfordshire OX20 1TR

press.princeton.edu

All Rights Reserved

Library of Congress Cataloging-in-Publication Data

Names: Sinclair, A. R. E. (Anthony Ronald Entrican), author. | Beyers, René, 1961– author.
Title: A place like no other : discovering the secrets of Serengeti / Anthony R. E. Sinclair with René Beyers.
Description: Princeton : Princeton University Press, 2021. | Includes bibliographical references and index.
Identifiers: LCCN 2020056501 (print) | LCCN 2020056502 (ebook) | ISBN 9780691222332 (hardback) | ISBN 9780691222349 (ebook)
Subjects: LCSH: Animal ecology—Tanzania—Serengeti National Park Region. | Ecosystem management—Tanzania—Serengeti National Park Region. | Biodiversity conservation—Tanzania—Serengeti National Park Region. | Serengeti National Park (Tanzania)
Classification: LCC QL337.T3 S56 2021 (print) | LCC QL337.T3 (ebook) | DDC 591.709678/27—dc23
LC record available at https://lccn.loc.gov/2020056501
LC ebook record available at https://lccn.loc.gov/2020056502

British Library Cataloging-in-Publication Data is available

Editorial: Alison Kalett and Whitney Rauenhorst
Production Editorial: Natalie Baan
Jacket Design: Layla Mac Rory
Production: Danielle Amatucci
Publicity: Kate Farquhar-Thomson and Sara Henning-Stout
Copyeditor: Steven Krauss

Jacket image: Zebras grazing in the Serengeti, Tanzania. Photo: James Smith / Alamy Stock Photo

This book has been composed in Arno

Printed on acid-free paper. ∞

Printed in the United States of America

10 9 8 7 6 5 4 3 2 1

In recognition of
Justin Hando
1951–2020
and
Markus Borner
1945–2020

Who, in different ways, saved the Serengeti in
its time of greatest need

CONTENTS

PREFACE

This is the narrative of how scientists discovered the rules that produced the unique features of the great Serengeti ecosystem and allowed it to persist. The story begins when I, Anthony Sinclair, first arrived in the Serengeti as a young student in 1965, and, seeing the magnificence of the place, decided to find out why it was so outstanding and what held it together and made it work. My first job, as an assistant to an Oxford University professor, was to answer the question of how so many migrant birds from the whole of Asia and Europe had managed to fit into the comparatively tiny area of East Africa for the winter—they seemed to be breaking all the rules of ecology. While engaged with this problem, I became aware that the large-mammal populations of Serengeti were increasing rapidly, especially the dominant herbivore species of buffalo, wildebeest, and elephant. I was asked to turn my attention to this problem. The increase raised a number of questions fundamental to ecology and essential for conservation.

I started with this simple question: Why were these species increasing? But this led to more profound questions and discoveries. Perhaps once a decade since then, there has been an important discovery concerning some new aspect that explains how the Serengeti works. These aspects are really all related to one fundamental principle—the principle of regulation—which governs not only Serengeti but also every other ecosystem of the world. It turned out there were seven of these aspects, which we called subprinciples, uncovered over a period of some 50 years (see chapters 3–10). The story of their elucidation is necessarily linear, and for the most part that is how the principles were uncovered—one discovery led to more questions, which led to the next discovery. The principle of regulation (chapter 3) was deduced in the 1960s and

'70s; bottom-up food limitation (chapter 4) in the 1980s; top-down predation (chapter 5) and migration (chapter 6) in the 1980s and '90s. Biodiversity and stability (chapter 7) are a combination of many aspects that came together in the 1990s and 2000s. Disturbance (chapter 8) was finally understood in the 2010s, while continuous change (chapter 9) became apparent to us from the paleontological work of others at Olduvai Gorge and our own work on vegetation in the 2000s. Our elucidation of multiple states (chapter 10) was first formulated in the 1990s, but it made more logical sense to place it at the end because it deals with complex interactions of other principles that had to be explained first. These chapters work toward a synthesis of the subprinciples in chapter 11. No doubt there are more processes to discover, and we mention, also in chapter 11, some of the interesting problems that will have to be addressed in the future.

Regulation and its subprinciples, then, form the basis of our understanding of what has gone wrong in the Serengeti, to explain the consequences of poaching, human intrusions, and the decline in drinking water for animals, for example (chapter 12). Building on this, we use the subprinciples to understand ecosystem malfunctions elsewhere in the world, using examples of each subprinciple (chapter 13). But we cannot leave the story there, in a state of doom and dejection. We have to ask, What can we, as conservationists, do about it? We address this in the final chapter. What we do is *rewilding*, the reconstitution of ecosystems as they would have been had humans not degraded them by overexploitation and disturbance.

This book is a description of the Serengeti, not only of what is there—the landscape, vegetation, and animals—but also of how it works. In much the same way, a description of the human body is not just the anatomy—the bones, muscles, skin, and blood—but also how it works, its physiology, its regulatory mechanisms that keep the body together, its behavior, the outside impacts that affect the body, and above all its pathology—what goes wrong with it. So also we describe the ecosystem; its anatomy is the community of species in their environment, but then we need to know how they all stay together—their regulation—what disturbs this combination of species, how it changes, and what goes wrong when parts of the anatomy and physiology are changed not just in this system but in other systems around the world, each with its own ecosystem pathology. By understanding this system, we can also understand other systems.

The combination of species in their environments, the anatomy of the ecosystem, is held together by a set of processes, which we will describe in the following chapters as the eight principles, most being elaborations on the nuances and complexities of the first principle, regulation. Each of these principles is known from other ecosystems in the world, but Serengeti is one of the few places (if not the only place) that is sufficiently described to show all of the principles in one ecosystem.

Serengeti is an easy place to observe, describe, and record. Its relatively flat terrain, the open vegetation of its plains and savanna, the mild climate, the ease of movement have all allowed scientists to record the changes in the ecosystem over many decades. Many ecologists have worked here since the 1950s. We report on what they have found in terms of the questions we ask about how the system works. We write this account for the broader human society to show why it is important to understand how an ecosystem works. If you don't know how it works, you don't know how to look after it. Bill Robichaud, now president of the Saola Foundation and a former student of mine, who works in Laos, described how a juvenile of an ungulate new to science was brought into captivity and fed a starchy gruel, on which it died. This was the first specimen of the saola, a primitive relative of the cow, ever seen alive, but because the people did not know how to look after it, it perished. The same argument applies to entire ecosystems. We will show how the wrong husbandry has resulted in the collapse of many ecosystems around the world. This is the legacy of past generations in fragile ecosystems on islands—even large ones such as Australia and New Zealand—as well as on continents. If only they knew a hundred and fifty years ago what we know now!

For ease of reading, we have put the scientists' names, backup justifications, and references (which in a scientific text would normally be in the main body) in the notes at the back of the book. Occasionally we refer to a scientist by name, when reference is necessary. Similarly, we have, wherever possible, left out the scientific names (unless no common name exists)—these can be found in the appendix, where we list all mammal and tree species mentioned in the text.

This is the account of how the different processes that drive the Serengeti ecosystem came to light over five decades. Many scientists have a study system to test their theories, and Serengeti is the one I used. In essence, it is a detective story showing how we came to understand the workings of the ecosystem, complete with theories, false starts, wrong directions, hard facts, remarkable luck, and surprising outcomes.

We hope to show that Serengeti and other large, protected areas act as the baselines against which we can monitor the tempo of human ecosystems to judge their stability and persistence. They act as the insurance policy for humankind, without which we cannot judge what we are doing to our world. Without these baselines we are blind to events. But many natural systems that have been degraded by human actions need to be repaired through rewilding so that they can act as baselines. The principles deduced from the study of Serengeti can be applied to most other ecosystems, both to understand their problems and to offer clues as to how to repair them. Serengeti shows that natural systems can repair themselves from major disturbances—even total collapse—if allowed time, protection, and perhaps a little help. That is the punchline from this book.

Anthony Sinclair and René Beyers
April 2021

ACKNOWLEDGMENTS

Over the years a myriad of people, too many to mention, have contributed to this story, made discoveries, and provided help. But they are not forgotten, and their contribution has been invaluable. A few to mention are Anne Sinclair, my wife, who has been with me for the whole 54 years of work, and long-suffering she has been. Also Simon Mduma, who started with me in 1988 and continues to this day, although now holding down the job of director-general of the Tanzania Wildlife Research Institute. Other researchers mentioned frequently are Mike Norton-Griffiths from Kenya, John Fryxell at the University of Guelph, Grant Hopcraft, now at the University of Glasgow, Ray Hilborn from the University of Washington, and Kristine Metzger, my data manager, now at the US Geological Survey, Albuquerque, all of whom worked closely with me. Students Ephraim Mwangomo, Ally Nkwabi, John Bukombe, John Mchetto, and field technicians Joseph Masoy and Stephen Makacha, among many others, contributed to the Serengeti Biodiversity Program.

Over that whole length of time, the Tanzania National Parks, the Serengeti park wardens, and the Tanzania Wildlife Research Institute have supported my work, allowing me to continue uninterrupted. Of particular importance in allowing this work were Tanzania National Parks Directors David Babu, Lota Melomari, Gerald Bigarubi, and Alan Kijazi; and Serengeti Chief Park Wardens David Babu, L. M. Ole Moirana, Bernard Maragesi, Justin Hando, and Martin Loiboki. Markus Borner, of the Frankfurt Zoological Society, played a pivotal role in keeping Serengeti going for some 40 years through his support of the Serengeti Wardens—simply put, without him and the enormous financial and material support of the Frankfurt Zoological Society, Serengeti would not exist today.

My work started with the help of Professor A. J. Cain, my supervisor Professor Niko Tinbergen, and Dr. Hugh Lamprey, the first director of the Serengeti Research Institute. Funding came from various sources, but the main one was the Canadian Natural Sciences and Engineering Research Council, which supported me for 44 years. Other funding came from the Royal Society (1965), NATO (1966–69), Texas A&M University (1970–73), the African Wildlife Foundation (1960s, 1970s), the Canadian International Development Agency (1970s), the Wildlife Conservation Society, New York (1980s, 1990s), and the Frankfurt Zoological Society (1980–2011). Some (but not all) of the important contributors to this story include Bernhard and Michael Grzimek (1950s), Murray Watson, Richard Bell, Dr. Hugh Lamprey, Hans Kruuk, George Schaller, Desmond Vesey-Fitzgerald, Hubert Braun, Hubert Hendrichs, Andrew Laurie, Terje Munthe, and Stephen Makacha (1960s); Peter Jarman, Patrick Duncan, Brian Bertram, Hendrik Hoeck, Dennis Herlocker (who produced the first and vital vegetation map), Hugo de Wit (who produced the first soils map), John Bunning, Helmut Epp, Robin Pellew, Jeremy Grimsdell, Jeanette Hanby, David Bygott, Bernard Gwamaka, and Joshua Moshi (1970s); Dianne Goodwin, Peter Arcese, Holly Dublin, Scott Creel, Neil Stronach, and Martyn Murray (1980s); Robert Fyumagwa, Sarah Durant, Craig Packer and his team, Sam McNaughton and Mike Anderson and their teams (1980s–2010s); Charles Mlingwa, Emmanual Gereta, and Greg Sharam (1990s, 2000s); and Tom Morrison, Sarah Cleaveland, and the whole epidemiology team at University of Glasgow (2010s). Karim Hirji, the first director of the Tanzania Wildlife Research Institute, made things possible in the 1980s, followed by Simon Mduma and Ernest Eblate. Chief Park Wardens David Babu (1970s, 1980s) and Justin Hando (1990s, 2000s) helped us survive during difficult times. In Arusha we relied on the invaluable logistical support of Bob and Kathy Gillis from the Canadian International Development Agency (1980s), Jo Driessen and Judith Jackson (2000s, 2010s), and Ben Jennings (2010s). While in Nairobi, we were looked after by High Court Judges J. Rudd and John and Stella Spry (1960s), Robert and Mary Ridley, Richard and Ruth Lloyd (1960s–1980s), Hugh and Ros Lamprey (1980s), and Mike and Ann Norton-

Griffiths (1977–2011). Over the years I received considerable encouragement and advice from my close colleagues and friends in Canada Charles Krebs and Judy Myers, in Australia Roger Pech, and in New Zealand Andrea Byrom—these last two introduced me to the concept of rewilding in 2010. Doug Houston, of the US National Parks Service, introduced me to the idea of photopoints from his work in Yellowstone National Park in 1978. Our colleagues at the Beaty Biodiversity Research Centre and the zoology department of the University of British Columbia have supported both René Beyers and me in many ways.

I have greatly appreciated the comradeship of my fellow scientists featured in the book *The Serengeti Rules* and the film of the same name: the narrator, Sean Carroll, as well as John Terborgh, Mary Power, Jim Estes, and Robert Paine. We are also indebted to Veerle Willaeys, René's wife, for her patience and support in this project.

Figures 2.1 and 4.2 were taken by Anne Sinclair. Figure 9.3A was provided by the Martin and Osa Johnson Safari Museum, Chanute, Kansas. Figure 10.1A was given to me as a photocopy by Mrs. Cynthia Downey in 1983, and published by Holly Dublin in 1991. Figure 10.1C was provided by Reto Bühler, Zurich, Switzerland, with the help of Joseph Ogutu, Stuttgart, Germany. Figure 14.1 is from US government ERTS satellite imagery recorded by N. H. Macleod from American University, Washington, DC. All other photographs were taken by the authors.

Finally, we thank Rob Pringle and an anonymous reviewer for constructive suggestions that have greatly improved the text. Our editor at Princeton University Press, Alison Kalett, provided the incentive to write this book. She and her assistant, Whitney Rauenhorst, production editor Natalie Baan, and copyeditor Steven Krauss have helped us throughout the publication process. Thank you.

A Place like No Other

1

Why Serengeti?

A WORLD HERITAGE

In 1972, the United Nations Conference on the Human Environment, in Stockholm, developed the Convention Concerning the Protection of the World Cultural and Natural Heritage.[1] Later in 1972, former US Secretary of the Interior Stewart Udall, a passionate advocate for the environment, who also served as US representative to the Stockholm conference, visited Serengeti. I met him there, took him around the park, and discussed the future. He recounted the events in Stockholm earlier that year: a debate had developed over how to choose World Heritage sites. As a first attempt, delegates were asked to retire for the evening and draw up a list of their top 10 preferred sites around the world. Next day all the lists were collated, and the one with most votes, the top of the list, was the Serengeti ecosystem. It was voted the most important natural area in the world.

Serengeti is outstanding for its biodiversity, its great migrations, and its iconic megafauna of large mammals. It is one of the last remaining relatively intact examples in the modern world of the last Ice Age, or Pleistocene. Why is Serengeti so different from any other place? Why is it regarded as the most important natural ecosystem in the world? All heritage sites are unique in their own ways, so why does this one stand out? We know from paleontology that the main aspects of Serengeti have been around for a long time, some four million years at least.[2] There must be a set of conditions and processes that create special features of

Serengeti and that result in its persistence over long periods. Over the past 50 years or so, a group of scientists has worked to elucidate these processes, which are governed by what we have termed *principles*. What makes Serengeti both outstanding and spectacular? What are the environmental features that allow a migration of so many animals? What determines the sizes of animal populations and the diversity of species that live there? Indeed, why does it have so many species? These are some of the questions we will consider here as we explore the biology of Serengeti. Using these principles, we can understand the problems facing Serengeti today, and what might happen to it in the future. These principles will also allow us to understand how problems in other areas of the world have developed and, finally, how we can repair them.

————

The Serengeti is defined by the area across which the wildebeest migrate. Serengeti is now a household name, the epitome of a wildlife spectacle in Pleistocene surroundings. Surprisingly, it has only recently come to be known thus. It was the lions that first attracted attention, in the 1920s—lions to be hunted by foreigners—and the wildebeest migration was completely unknown. The Serengeti plains were the place to go for the grandest black-maned lions in the world, and there were lots of them to shoot. It was not until the Germans Bernhard Grzimek and his son, Michael, flew their plane over the Serengeti in the late 1950s to document the great migration in their film (and book of the same name) *Serengeti Shall Not Die* that the world first became aware of the phenomenon.[3] The Serengeti is significant because it supports one of the last remaining migrations of large mammals in a relatively unchanged state from the time of the hunter-gatherers, long before the agricultural development that gradually emerged in the 1600s from the Congo, far outside Serengeti, and before the impacts of the modern economic world were felt. It is also a place of singular beauty and remarkable biodiversity: it supports more large mammal species than any other place in the world, and almost as many bird species as the whole of Europe. Despite its relatively undisturbed state, the ecology of the

Serengeti has changed over the past century, and these changes high-light its fragility and sensitivity to climate and human impacts.

Serengeti is a place where biologists can observe nature more easily than most. Its combination of open plains and savanna allows access to most of the area. The large animals are readily observable. One can describe their ecology and behavior using only binoculars. Their populations can be counted accurately. Because of the many decades biologists have been studying the Serengeti, we now understand the causes underlying the huge changes that occurred in the ecosystem both in the distant past and during the past century. By now nearly everyone knows that human impacts on nature are becoming ever more severe, and Serengeti has become a case study documenting these impacts. Long-term studies have shown how political, economic, and social events have driven the ecological changes.[4] Serengeti has become a vital source of information for science on how ecosystems work and how they respond to pressures.

————

My involvement in Serengeti began on July 1, 1965, driving south from Nairobi through the Maasai Mara Park to Banagi, the research head-quarters of a small band of biologists in the center of Serengeti National Park. My job was to record the bird migrations from Asia, as an assistant to Professor A. J. Cain of Oxford University. He had projects elsewhere in Africa and left me to it for three months. I was given a small round-house to live in. My first morning, at dawn, I accompanied the park warden's driver while he read all the rain gauges scattered around the park, a job that was done at the end of each month. It was the first of three days of rain-gauge reading, and in that time we covered the whole of Serengeti, some 20,000 square kilometers.

At the end of those three days, I had seen the Serengeti as few non-natives had ever done. In the past I had seen something of East Africa, having been raised there, and had visited various game parks. But nothing had prepared me for this experience of wildlife in vast numbers, the extraordinary migrations, the sheer diversity of animals and vegetation, and the spectacular landscapes. There had to be a reason why Serengeti was such an outstanding place, and I decided to find out, to discover the conditions, the processes and underlying principles that made Serengeti the way it is. But this system, though unique, shares many features with other ecosystems in the world, making it a model for understanding ecological processes.

The principles are useful not just to explain how Serengeti itself works, in its unusual, even aberrant form. In understanding this ecosystem, we can begin to make sense of all other systems by recognizing how they differ from Serengeti; they are the other side of the coin. This book recounts the history of how these principles were discovered.

———

Perhaps the best way to begin is with a brief description of the Serengeti ecosystem. Its special geographic features determine its physical environment, climate, water relations, and habitats. Together, these create the conditions that make possible the great migration of wildebeest and other species.

The Serengeti-Mara ecosystem is an area of approximately 25,000 square kilometers on the border of Tanzania and Kenya in East Africa, and its extent is defined by the movements of the migratory wildebeest. This includes many political administrations. The main ones in Tanzania are the Serengeti National Park (SNP) itself, and the Ngorongoro Conservation Area (NCA), which lies east of the park and includes half of the Serengeti plains. North of the NCA is the district of Loliondo. The Maasai Mara Reserve is the main Kenyan administration. This area holds the vital dry-season grazing and water supplies for the migration. South and west of SNP are small game reserves, such as Maswa, Grumeti, and Ikorongo (figure 1.1).

Most of the ecosystem consists of a flat or rolling landscape highly dissected by small seasonal streams that flow into a few major rivers. It is part of the high plateau of interior East Africa. This gentle aspect slopes from the edge of the Gregory Rift in the east down to Lake Victoria in the west, so that all the rivers (except the Olduvai, on the plains)

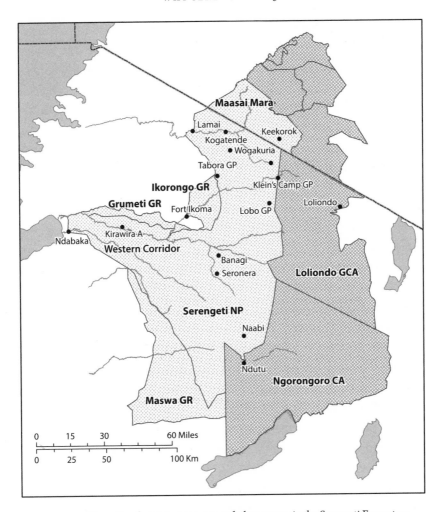

FIGURE 1.1. The main administrative areas and place names in the Serengeti Ecosystem.
NP = National Park, GR = Game Reserve, CA = Conservation Area, GCA = Game
Controlled Area, GP = Guard Post.

flow west. The highest part of the plains is at an altitude of 1,800 meters,
while Speke Gulf in the west is at 1,200 meters.

There are three major rivers, the most important being the Mara,
which originates in the montane forests of the Mau Highlands of Kenya.
It has until recently flowed year round (see chapter 12), providing the
main water supply for the great herds of migrating animals in the dry

season. It flows through the Mara Reserve of Kenya and northern Serengeti, and eventually flows west through the huge Musiara swamp into Lake Victoria at Musoma. The two other rivers are the Grumeti, which originates in the highlands of northeastern Serengeti, and the Mbalageti, which originates on the Serengeti plains. Both are seasonal rivers with only pools remaining in the dry season. Two more rivers originate in southern Serengeti, the Simiyu and the Duma, but only their upper reaches lie within the Serengeti before they flow through agricultural land to Speke Gulf. All other rivers dry out except for a few springs that seep from the base of hills.

Steep, rocky hills occur along the eastern boundary of SNP and between the Grumeti and Mbalageti rivers in the west, forming a backbone to the corridor between the rivers. The Nyaraboro Plateau, with a high (300-meter) escarpment, occurs in the southwest. Because of the generally higher elevation in the east, the hills in Loliondo and the northeast of SNP reach 2,000 meters.

The ecosystem is effectively self-contained, enclosed by natural boundaries on all but one side. The eastern boundary is formed by the escarpment of the Gregory Rift and the base of the Crater Highlands. The south is bounded by the edge of the Serengeti plains and in Maswa by the appearance of numerous kopjes (rocky outcrops). In the west, the corridor, which is largely an alluvial plain formed by the rivers, is bounded on both its south and north by higher ground—now agricultural land—and by Speke Gulf. The west side of the northern extension of Serengeti to the Kenya border is an artificial boundary set by agriculture. Within Kenya, the Mara Reserve is bounded by the Isuria escarpment, the Loita plains, and the Loita hills (figure 1.2).

———

The Serengeti ecosystem is (very roughly) a square, with the treeless plains covering the bottom right quarter, about 5,000 square kilometers. They were formed by dust deposits from the volcanoes of the Crater Highlands 4 million years ago. To understand how this happened, we have to go back to the Miocene, some 14 million years ago, when eastern

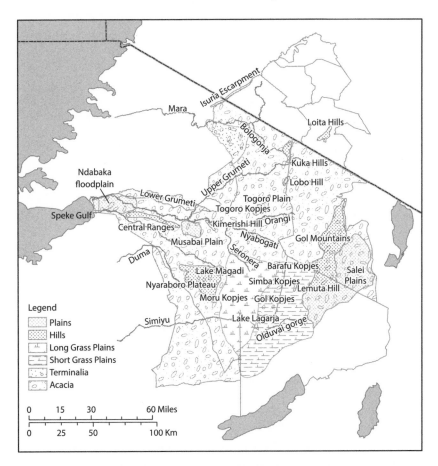

FIGURE 1.2. Topography and habitats of the Serengeti ecosystem.

Africa began to split apart due to plate tectonics—the same process that split Africa from South America starting 100 million years ago. Africa is still breaking apart and in a few million years will be two continents. The split is developing down a rift, the Great Rift Valley, from the Dead Sea in the Near East through Ethiopia to East Africa. In East Africa this rift splits into two arms (figure 1.3). The western arm, called the Albertine Rift, runs along the western borders of Uganda, Tanzania, and Mozambique. Within it lie the deep lakes Albert, Tanganyika, and Malawi. The eastern arm, the Gregory Rift, runs through the middle of Kenya and Tanzania. The edges of each rift are uplifted so that the land between

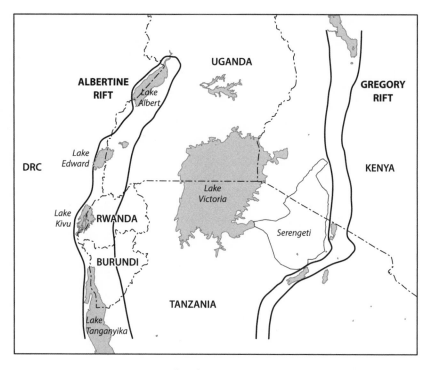

FIGURE 1.3. The rifts and lakes of East Africa.

the two forms a shallow basin. Lake Victoria is impounded in this basin, essentially a vast and shallow puddle only about 65 meters deep at its greatest depth. At 65,000 square kilometers, it is huge, the largest lake in Africa and the third largest in the world after Lake Superior and the Caspian Sea (which is in fact a lake). It is some 200 kilometers across both west and north.

Over time the rifting resulted in volcanoes, and the Crater Highlands of Ngorongoro, made up of several different volcanoes, developed. These were very active four million years ago, during the Pliocene period. Prevailing winds from the Indian Ocean in the east blew the dust westward; it settled out, deeper close to the volcanoes and gradually becoming shallower as it was blown farther west. Eventually this dust hardened into an impenetrable layer of calcium, a hardpan (called calcareous tuff) close to the surface. Tree roots cannot penetrate this, and so cannot take hold, but grasses and herbs, especially creeping ones, can establish themselves.

The volcanic soil was rich in nutrients, especially nitrogen, phosphorous, and calcium, and the creeping plants became highly nutritious. The volcanoes are quiet now, but one, Oldonyo Lengai, is still active, with significant dust fallout in 1966 and 2007. The Ngorongoro Crater itself is a caldera, the sunken base of an old volcano.

Far enough west from the volcanoes, the hardpan becomes sufficiently thin and broken up that trees can get their roots through and establish themselves. This produces something unusual in ecology, a very narrow boundary between trees on one side and no trees on the other. This boundary, only a few meters wide, runs along the northern, southern, and western sides of the plains. Seronera, the park headquarters, lies in the corner of the western and northern boundaries. The geology from the edge of the plains westward and toward the center of the park is largely ancient granite with sediments from hills and rivers.[5]

Climate

There are two special climatic features that determine the Serengeti environment. First, the Crater Highlands in the southeast are sufficiently high (2,400 meters) that they impede the prevailing winds from the Indian Ocean, causing a rain shadow on their western side, where precipitation is scarce. The far-eastern Serengeti plains, therefore, are semi-arid, receiving only 500 millimeters of rain per year. Sand dunes are gradually moving across that region. The second important feature is Lake Victoria, in the west. This lake is so large that it creates its own weather system; rainstorms develop over the lake and affect the west and northwest of the ecosystem, even during the dry season. So there is a wet northwestern region and a dry southeastern region, producing a marked gradient in rainfall. It is this gradient that drives the migration and ultimately determines the working of the entire ecosystem.

Serengeti is a highly seasonal environment, with rainfall being the major influence. Rain is determined by the position of the sun. Serengeti is near the equator, lying just two degrees south. The sun passes over it on its way south in September and on its way north in March. About six weeks after the sun passes, a band of heated air, the Intertropical

Convergence Zone, follows the sun and draws in wet air from the Indian Ocean. As a result, there are two wet seasons, a shorter one in November and December and a longer one from March to June. Both seasons are variable, the shorter one often absent, but sometimes the two merge to form one long wet season—as occurs when there is a strong El Niño event in the Pacific Ocean. There is a long dry season from July to October. July is the driest month, giving way to storms that become more frequent in subsequent months in the northwest of the system. In addition, Lake Victoria contributes rain from storms generated by heating over the lake.

Habitats

We can think of Serengeti as having three major habitats, which, by coincidence, lie along the rainfall gradient (figure 1.2). First, there are the plains in the southeast, which as already mentioned are formed by a calcareous layer under the soil derived from volcanoes long ago. This layer prevents trees from growing, and so the plains are open grassland. However, there is also a gradient of grassland types within the plains. The far-eastern plains are very short grasslands, composed roughly of 40 percent grasses and sedges, 40 percent small flowering herbs, and the rest bare ground. All these plants are heavily grazed, though they grow close to the ground as protection from being eaten. In the middle of the plains, the grasses are longer (30 centimeters), mixed also with flowering shrubs; these are the intermediate plains. To the west and north lie the long-grass plains, where the grasses grow up to a meter. The Salai plains form the northern half of the eastern grasslands in the Loliondo district. They are very dry and contain sand dunes, most of them covered with tussock grass, except for some that are still moving.

The second habitat, the acacia savanna, starts abruptly at the edge of the plains as the effect of the volcanic soil disappears. The majority of the savanna is formed by many species of African acacias, often as single-species stands. Each has its own preferred position along a gradient of drainage. The gently rolling landscape results in well-drained ridgetops with sandy soils. The soil type changes progressively down the slope

FIGURE 1.4. The short-grass plains at Gol Kopjes with wildebeest, May 1971.

until poorly drained and even waterlogged silt soils are found at the bottom, near rivers and drainage lines. The African acacia species lie along this gradient—umbrella trees (*Vachellia [=Acacia] tortilis*) prefer the tops, stink-bark acacia (*V. robusta*) prefers mid-slope, while yellow-fever trees (*V. xanthophloea*) like wet soils near rivers, and gall-acacias (*V. depranolobium, V. seyal*) live in waterlogged, swampy soils. These are just a few of the many tree species that occur here. The grass layer is composed of species similar to those on the long-grass plains, but under trees there are other species together with many flowering herbs. The grass layer in the savanna is much richer in species than on the plains. Along riverbanks and around kopjes are shrubs that form thickets—favorite hiding places for predators (figures 1.4, 1.5, 1.6).

The third major habitat is the broad-leaved woodland of the far northwest, composed mainly of trees of the genera *Terminalia* and *Combretum*. The soils are derived from granite and, though poor in nutrients, support tall grasses (up to 2 meters) and many different species of shrubs.

There are also special habitats. The western end of the corridor was, until a few thousand years ago, under Lake Victoria; it is now a flat

FIGURE 1.5. Savanna with dense stands of stink-bark acacia, facing west to Lobo Kopjes in background, November 2005.

FIGURE 1.6. Broad-leaved *Terminalia* woodland with tall *Hyparrhenia filipendula* grassland in northwest Serengeti, March 2005.

floodplain, the Ndabaka, with alternating clay soils and sandy ridges (old beaches). It floods during the rains. The hills in the center of the park, on the Nyaraboro plateau, and along the eastern boundary are stony or have thin soils. Species of *Combretum* are found on lower slopes, but a mixture of small acacias and shrubs grow higher up. On the higher hills of the northeast, such as Kuka Hill, the elevation results in montane forest, a relic of that on the Crater Highlands and the Loita hills in Kenya. Most of this forest has been destroyed by fire over the past century, and the hills are now covered by grassland, but small patches of forest remain in gullies where fire cannot reach.

The Mara River supports riverine forest, which is an extension of the montane forest downstream from the highlands in Kenya, where the river has its source. This is closed-canopy forest maintained by a high water table from the river. It can be half a kilometer wide but more usually is 50 meters or less in width. There is evidence it was much more extensive in past centuries. The Grumeti and Mbalageti rivers along the western corridor also support forest, but this is forest of a completely different type. It is a subset of the lowland Congo forest and has almost no overlap in tree species with the Mara montane forest.[6] The Serengeti thus supports two major forest types, only 50 kilometers apart.

The rocky outcrops, or kopjes, found in the eastern half of the eco-system are granite intrusions and are surrounded by a matrix of volcanic rocks. These small islands of rock support dense shrubs and broad-leaved trees such as marula (*Sclerocarya birrea*) and fig trees (*Ficus* species). Vegetation is scarce on the kopjes of the eastern plains but lush on those at Moru, an area of large kopjes at the western edge of the plains in the south of the park. Kopjes are small in area but are an important and special habitat for animals.

The Animals

The most numerous big mammals in the Serengeti are the ungulates. These are large hoofed animals such as antelopes, zebra, buffalo, giraffe, rhinoceros, and hippopotamus. Every year during the great migration about 1.5 million wildebeest, together with 200,000 zebra and half a million Thomson's gazelle move around the system—over 2 million animals in all (figure 1.7). They all converge on the plains in the wet season because that is where the best food is. The grasses of the plains have the highest protein in the whole of Serengeti, but calcium and phosphorus are also high. The animals move around the plains following the rainstorms and the growth pattern of the grasses.[7] The three migrant grazers, however, stick to their own kind, with only a small overlap in their distributions, each taking advantage of the different heights of grass. As the plains turn green with the first rains, usually around December, the Thomson's gazelle arrive first, feeding on the short new growth. Then, as the grass grows a little taller, say 15 centimeters, the wildebeest arrive and displace the gazelle, which now move farther east to the short-grass plains. Eventually the zebra arrive, and they confine themselves largely to the intermediate grass plains. This sequence of zebra on the western plains, wildebeest in the middle, and gazelle in the east moves farther east in lockstep as the wet season progresses and the plains become wet and waterlogged in April. They all move back west in the reverse order, with zebra going first, when the plains dry out in May or June. However, in January and February there is quite often a dry period (between the two rains), and in that case the whole sequence moves south and southwest into the Maswa Game Reserve, a 2,200-square-kilometer area bordering the Serengeti National Park. The reserve is a vital retreat at this time of year. One other species also migrates. This is the eland, the largest of the antelopes (males can stand 1.6 meters at the shoulder and weigh more than 600 kilograms). Some 15,000 of them move onto the plains in the wet season, feeding on herbs and also grasses when the latter are green (figure 1.7).

June sees the migration moving west and north. It is at this time of year that the herds graze the long grasslands. They move slowly—both because it takes time to graze long grass and because they are wary of predators—and so bunch up and form the dense masses that have become famous from photographs. Wildebeest dictate the movements of the other species. They eat down the grass and provide a food niche of short grass for the gazelle that follow behind.[8] Zebra are often found close to the wildebeest, probably because of safety, but they must stay

FIGURE 1.7. Wildebeest migration, Seronera, June 2004.

in front because they need a greater bulk of food than the wildebeest. A picture of the great herds would show a front fringe of zebra, a mass of wildebeest, and then a dispersed scattering of gazelle behind.

Once the herds reach the woodlands, this pattern breaks up and smaller groups of wildebeest and zebra make their way west and north during July and August. Thomson's gazelle stay behind in the central woodlands, and by the later dry season there is little overlap with the other two species. By the end of the dry season (September, October), the wildebeest are in two major groups, one in the corridor between the Grumeti and Mbalageti, the other in the northwest of Serengeti and in the Mara Reserve of Kenya. Eland move north to the *Terminalia* woodlands, where they feed on the more abundant shrubs.

The beginning of the rains, in November, brings the migrants south and east again toward the edge of the plains. However, the rains usually begin with scattered thunderstorms, which cause the herds to spread over most of the woodlands in search of local patches of green food.

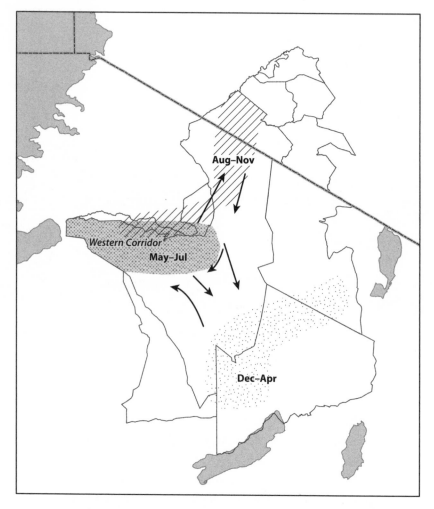

FIGURE 1.8. The seasonal distribution of the migratory wildebeest and zebra. In the wet season (December–April) they are on the plains; in the dry season (August–November) they are in the corridor and north into Kenya.

They are at their most dispersed at this time, only congregating again when the rains become more consistent and the herds can once again move onto the plains (figure 1.8).

———

The distinction between those we call *migrants* and those we call *residents* is more of scale than of absolute differences. All species move with the seasons, but whereas the migrants move hundreds of kilometers, the residents usually move only a few.

The commonest resident antelope is impala, the quintessential animal of Africa. Paradoxically, it is the only one of its kind; it has no close relatives. Impala live in herds of 10 to 200, all females and youngsters, and one male. All other males are forced out to live in large bachelor groups.[9] Impala live wherever there is savanna, feeding on shrubs, herbs, and green grass, never venturing onto the plains.

Topi and kongoni are both close relatives of wildebeest, all being members of the family called Alcelaphinae, and they all eat grass almost exclusively. Both species are sedentary, topi living on wet grasslands in large groups of sometimes several thousand in the western corridor, while kongoni keep to the eastern woodlands and long-grass plains, where conditions are much drier. There are no kongoni in the corridor. Both species overlap in the center and northern woodlands, living together in small groups.

There are several species of antelope that live in more restricted habitats. Water-loving antelope include waterbuck (at 270 kilograms and more, a male waterbuck can be larger than a wildebeest) and the smaller common reedbuck, which is abundant in all tall grassland including the plains but rarely seen except along rivers. The rare mountain reedbuck is found only on the tops of the highest hills. Bush or forest antelope include bushbuck wherever there is thicket along rivers and greater kudu in the kopjes of Maswa. Lesser kudu are found in the montane forests of the Loita hills, which are not part of our system, but it is possible they are in the forests of the Serengeti highlands in Loliondo, yet to be discovered.

On the Salai Plains we find the desert-adapted beisa oryx. These are rare and may now be extinct in our system. Related to these is the roan antelope. They prefer, strangely, the low-nutrient granitic savanna typified by broad-leaved woodland. They were once abundant over the whole of northern Serengeti and Mara, some associated with the central hills such as Banagi, but they have since disappeared. One small group is being nurtured in the Grumeti Reserve and another in the Maswa Reserve. Another species that prefers the broad-leaved woodland is oribi, a small antelope that lives singly or in small groups,[10] mostly in the far northwest. Other small solitary antelopes include grey duiker in similar broad-leaved habitat, and steinbuck, which prefers dry acacia savanna.

African buffalo were also abundant, reaching 70,000 in the mid-1970s, but after the border closure with Kenya in 1977 rampant poaching reduced their numbers to 20,000 by the 1990s. They have now returned to original densities except in the northwest, where poaching continues.[11] Buffalo live in the savanna in herds of up to 1,500; they feed on long grass and need daily access to water. Giraffe are also ubiquitous in the savanna. Numbers are not well known, although we think there are about 8,000. Megan Strauss considers that they may be declining due to poaching.[12]

Rhino were once common over the entire Serengeti, including Olduvai Gorge. But again, beginning in 1977, the same poaching effectively exterminated the rhino and only a few remain in the south and in northern Serengeti and the Mara Reserve. Careful guarding of these few has kept the population alive since 1990, but they have not increased much.

———

Serengeti was first famous for its lions; hunting expeditions of foreigners would camp at the edge of the plains on the Seronera River because there were so many lions in the vicinity. It was the extermination of lions that led to the first small part of Serengeti becoming a protected area in 1930. Only in 1951 was the larger park legally demarcated. The Seronera River is still the best place to observe lions, although they range over the whole ecosystem.[13] Leopards live throughout the savanna and can also be found around the kopjes in the long grassland wherever there is dense vegetation. They are always secretive and difficult to see. Cheetahs are seen most frequently on the plains, but they occur in the savanna as well. There are also several small cats—serval, caracal, wildcat—but they are not often seen.

Hyenas are the most abundant carnivores in the system, some 7,000 of them.[14] They live in groups on the plains and hunt large ungulates together. They follow the migrating herds around the plains and some way into the woodlands, perhaps 50 kilometers. Once the herds get beyond this, hyenas that live on the plains do not follow. There are hyenas living throughout the savanna, hills, kopjes, and even forest; these are usually solitary and scavenge or feed on small animals. Hyenas that live in packs are also found on the larger grasslands within the savanna, such as those at Musabi, Togoro, the Mara Reserve, and the Loita plains in Kenya.

Several species of the dog family are also found in the Serengeti. The best known are wild dogs, which were once found throughout the savanna and plains. Strangely, European hunters and wardens disliked them and shot as many as they could (as they did hyena)—despite the area's being protected. Then, starting in the late 1960s, wild dog numbers within the ecosystem began a decline, which was linked to the increase in larger predators and the spread of disease from domestic dogs (described in chapters 5 and 12). The decline ended with their disappearance in 1992. A few packs have since been discovered on the fareastern edge of the ecosystem; they had probably always been there, undiscovered, but may have come in from elsewhere. Some of these animals have since been reintroduced to the western corridor and are thriving. There are four other dog types: the golden wolf (previously known as the golden jackal) is found mainly on the short-grass plains and around Ndutu; the black-backed jackal is common in the savanna; and the side-striped jackal, which is solitary, is seen only occasionally in both long-grass plains and savanna. The bat-eared fox, closely related to dogs, specializes in eating dung beetles—its large ears allow it to hear beetle larvae underground. It lives in holes throughout long-grass plains and the open areas of savanna.

The Serengeti also hosts many species of small carnivores, most of which feed on rodents and insects. These include members of the weasels (family Mustelidae) such as the zorilla and honey badger; six species of mongooses (family Herpestidae); and the civet, common genet, and spotted genet (family Viverridae).

The ground pangolin (order Pholidota) also occurs in low numbers. Being nocturnal, it is hardly ever seen, but since it is one of the rarest (and most threatened) mammals in the world, it is of special interest.

———

Recent taxonomy using comparisons of DNA (genomes) has revealed an ancient group of mammals that evolved when Africa was cut off as a great island, some 100 million years ago, at the dawn of the age of mammals. They are called *Afrotheria*, and they include the elephants, hyraxes, aardvark, and elephant shrews: a truly African group.[15]

Bush elephants (those that occur in the Congo forests are now considered a separate species of the smaller forest elephant) occur in savanna and long-grass plains and number about 3,000. They could be considered as migrants. There are two and possibly three populations, a northern group based on the Mara River and its forests in the dry season—including both the Kenya and Serengeti portions—and a southern group based on the Duma and Simiyu rivers and rivers in Maswa. In recent years, a third group has appeared in the western end of the corridor based on the Grumeti and Mbalageti rivers. The northern group moves south about 150 kilometers toward Seronera in the rains. The southern group moves east about 50 kilometers to Moru, and even across the plains to Lake Lagarja when it is wet. The western group appears to remain more sedentary in the corridor.

Hyraxes are rabbit-sized animals that live in kopjes and forest trees. Two species—rock hyrax, which feeds on the ground, and bush hyrax, which feeds in trees—live together in the crevasses of kopjes throughout the savanna and a few kopjes on the edge of the plains where there is enough shrubbery. The third species, the tree hyrax, lives only on the Mara River in the holes of large riverine trees.

Perhaps the most unlikely relative of the elephant is the aardvark. With its strange, tubular teeth, long snout and tongue, and long digging claws, it is supremely adapted to opening up mounds to eat ants and termites. It occurs everywhere there are termite mounds, digging into these from the side to access the nests and using its long tongue to lick up the insects. It is nocturnal, solitary, and rarely seen. But its influence on the ecosystem is large, for many species use the vacant termite holes as their houses.

———

Among other animals, biodiversity is very high in some groups. More than 600 species of birds flourish because of the high diversity of habitats, and many are influenced by the wildebeest migration. For tourists, birds are one of the more obvious features of the Serengeti. In contrast, rodents are not obvious, but they are also diverse; some 30 species occur, and many of the small carnivores and birds of prey depend on them.

Insects are not yet well described, but we have found approximately 180 species of butterflies, 100 species of dung beetles, and 70 species of grasshoppers. The insect fauna support the majority of the bird species. In contrast, both the reptile and amphibian fauna are not diverse. There are only a dozen species of lizards, 20 or so of snakes, and 30 of amphibians.

People

Humans have been involved in the Serengeti ecosystem for probably the entire length of human evolution. As described earlier, relatives of modern human beings lived around the lake now called Olbalbal, at the foot of the Crater Highlands, 1 million years ago. In the millennia since, humans likely persisted first as scavengers following the great migrations, and then as hunters. Artifacts found at Olduvai, Lake Lagarja, show they lived and hunted around the lake. A hand axe that could have been 100,000 years old was found on a ridgetop in the *Terminalia* woodlands. Other artifacts eroding from the banks of the Mara River at Kogatende could also be tens or hundreds of thousands of years old. Clearly humans and their earlier relatives have been in the system for a long time. However, we have no knowledge of what impact they may have had in shaping the system; it was simply too long ago. We may surmise that their populations were so low that their predatory impacts on the ungulates would have been quite small.

The most important impact of early humans was the use of fire. There is a scrap of evidence to suggest that fire may have been present 1 million years ago, but this could have occurred through lightning strikes.[16] Nevertheless, humans used fire early on, meaning that the ecosystem has been subjected to burning for hundreds of thousands of years; humans may be the reason why savanna vegetation has evolved to become fire adapted (see chapter 8). Large savanna trees such as *Acacia* species and broad-leaved *Terminalia* and *Combretum* can withstand burning. Even hot fires that scorch leaves merely result in shedding of dead leaves and regrowth of new ones. Young trees (less than 2 meters tall) within the reach of fires are usually scorched back to ground level. But they do not die; most have large rootstocks containing a substantial food reserve, which the plant uses to sprout again in the next rainy season. Eventually, some of these small trees manage to escape fire and grow into mature trees to maintain the savanna. Some shrubs and bushes, such as *Ormocarpum*, have evolved corky bark that withstands burning. Some plants take refuge within the dense undergrowth of other species. The candelabra euphorbia tree (*Euphorbia ingens*), for example, cannot grow in open grassland because its fleshy, mesophyllic stems are rapidly destroyed by fire. However, it starts life by growing within the lower thicket at the base of the umbrella acacia (*V. tortilis*). The young candelabra grows straight up through the top of the acacia, then branches out to form the characteristic mature tree, while killing the acacia, probably to improve root competition. Other small flowering herbs also take refuge in thickets where fire cannot penetrate. Thus does fire promulgated by humans shape the Serengeti plant community.

Humans, of course, have a pervading and all-encompassing impact on the ecosystem through their influence on global climate change (see chapter 13). These are relatively recent events. In this tropical system just off the equator, climate change manifests more by altering rainfall than temperature. Although increases in temperature have been detected,[17] they are small. Much more important is the prospect of major fluctuations in rainfall. As elsewhere in the world, the highs and lows are becoming more exaggerated, in this case creating longer dry periods and greater floods. The worst drought of the twentieth century occurred in 1993. The greatest floods, exceeding all records, took place in 2019–2020. These events are strongly influenced by what is known as the Indian Ocean Dipole, the heating system in the Indian Ocean that determines the rainfall in Africa and elsewhere. At the same time, the El Niño Southern Oscillation, driven by heating of the western Pacific Ocean (similar to that of the Indian Ocean), is showing more frequent and more extreme events, which affect nearly all components of the Serengeti ecosystem[18] (see chapter 8). These impacts on the Serengeti cannot be avoided.

In contrast to these long-term natural impacts, the shorter-term impacts of humans during most of the nineteenth and twentieth centuries have been largely confined to small-scale bushmeat hunting,[19] with the exception of the ivory trade (see chapter 4). The considerable historical evidence[20] shows that while hunter-gatherers were present in small numbers over the past 300 years, there was no agricultural settlement in the main part of the Serengeti ecosystem, the savanna, because of the tsetse fly. Agricultural tribes arrived from west of Lake Victoria in the sixteenth century but did not inhabit the eastern shore until the nineteenth century. Even then, they remained far to the west of the present Serengeti ecosystem. Only toward the end of the nineteenth century did agriculture arise in a narrow strip north of the corridor (and hence not in the ecosystem) along the Bunda–Ikoma road, to which it is still limited today. While this strip is suitable for agriculture, the soil and water in the area to the south, inside the Serengeti, will not support crops. Similarly, to the south, the Wasukuma people are limited to higher ground outside the current Serengeti boundaries, just as they were in the nineteenth century. To the northwest, the Wakuria people were limited to areas close to what is now Musoma, and nowhere near Serengeti, moving east only in the 1950s but remaining outside the tsetse belt in Serengeti. To the east, Maasai pastoralists were recent arrivals in the mid-1800s, and used only the far-eastern plains to avoid both the tsetse and the wildebeest. Only later, in the 1930s, did they move west toward Moru in the dry season for temporary grazing but retreated when the wildebeest returned. Overall, modern humans have had little impact on the Serengeti savanna in recent centuries.

Because of these circumstances, the Serengeti ecosystem becomes one of the most suitable areas in the world to act as a baseline for assessing the impact of humans on their own habitats. Agriculture being the single greatest disturbance to ecosystems, there is a concern that it could ultimately destabilize the systems on which humans depend.

Thus, for the benefit of future generations we would be well advised to both detect and monitor these human-induced perturbations. To detect disturbance by humans it is necessary to have a number of areas without that human impact in order to compare them with areas of agricultural development. These are the controls, the baselines that act as reference points. They are in fact the insurance policy for humanity in the future. It is for this purpose that the Serengeti has been studied and described here. We are not suggesting that Serengeti (or indeed any other baseline ecosystem around the world) is or even should be a pristine environment, without humans; probably no such place exists. Serengeti has been influenced by humans since the dawn of humankind. But it is the modern impacts of agriculture, forestry, and urbanization that we want to measure by comparing with areas free of those impacts.

———

This is just a rough outline of the immense richness of the Serengeti ecosystem. The variety of species is enormous, encouraged by the diverse habitats and the impact of the great migration itself. This is the Serengeti as it now stands. It has arrived in this form through a long series of changes caused by events over the past 200 years and more. These events were crucial in allowing us to understand how the Serengeti worked. The first of these formative events was the appearance of the viral disease rinderpest. The effects of this disease proved to be fundamental not only for understanding Serengeti, but also for shaping the ecology of the whole of Africa, probably for centuries.

2

The Discovery of Rinderpest

It all began in August 1965. I was trying to find out how overwintering birds migrating from Asia were able to fit into the much smaller area of eastern Africa.[1] The previous May, my fellow student Richard Bell had conducted a count of large ungulates in the Serengeti ecosystem, and he had persuaded me to help him count the animals in aerial photographs. He had shown that buffalo, wildebeest, and elephant were all rapidly increasing in number, and it was not long after this that I was asked to come back and study the changes in these populations. And so I began to work on buffalo in October 1966 and, as it turned out, the wildebeest as well because I soon realized that to understand buffalo population changes I also had to understand those of wildebeest. The first research question, therefore, was: What was causing these increases?

Possible answers included an environment that was improving because it was providing more food, perhaps a decreasing number of natural predators, less poaching due to the protection provided by the newly created Serengeti National Park (established a mere 14 years earlier, in 1951), or perhaps fewer or less severe diseases of the animals. Rain is the most important aspect of the Serengeti environment because it determines the amount of grass food that buffalo eat. Rainfall records had been kept since the 1930s; there was no obvious trend, especially not an improving one. Scientists had little idea of predator numbers at that time, but there was no obvious decline (in fact, as I will show later, numbers were increasing). A response to increased protection provided by antipoaching patrols was a good candidate, however.

I was aware of the rinderpest issue from the start. Rinderpest, a viral disease affecting cattle and wildlife, was regarded as the single most pressing problem for animal husbandry in Africa after the Second World War, yet little was known about it.

Initially it killed huge numbers of cattle and wildlife, which suggested it had not been in Africa for long. Where had it come from? We now know that it arrived in Africa in 1887 and created one of the world's greatest pandemics. The result was one of the most serious socioecological disasters in human history, on a par with the bubonic plague in Europe in the 1340s, smallpox in the New World after 1492, and presently COVID-19. Rinderpest is a viral disease of cattle that occurs naturally in Asia. It was introduced to Ethiopia in 1887 via cattle brought from India by Italian invaders; there is no evidence that it was ever in Africa prior to this event. It spread to West Africa, then south through East Africa in 1890, to Malawi in 1892, reaching the Cape by 1896. The resulting pandemic killed over 95 percent of cattle throughout Africa, often destroying complete herds. Famines then decimated the human population two to three years later in many parts of Africa. By 1892, widespread famine was ravaging Ethiopia, Somalia, southern Sudan, and eastern Africa.[2] At least three-quarters of a million people died in Tanzania in the mid-1890s.[3] In 1951, E. B. Worthington, a senior scientist at the Food and Agriculture Organization of the United Nations, recommended that research on wildlife be initiated in East Africa, solely with a view to solving the rinderpest problem—no wildlife research had been conducted in East Africa prior to this.[4] Subsequently, virologists—English veterinary scientist Walter Plowright in particular—set up a program to monitor and then reduce the incidence of disease with a cattle-vaccination program, starting in the mid-1950s.

Clear evidence that rinderpest swept through Serengeti in 1890 was reported by Austrian explorer and geographer Oskar Baumann when he passed through Ngorongoro and Serengeti.[5] He visited Maasai villages where all of the cattle had died and the people were starving. They pleaded for some of Baumann's cattle. Villagers ate anything available, scavenging the remains of carcasses, even the skin, bones, and horns of cattle. Baumann and his porters gave the starving people as much food as they could, but this brought in ever greater numbers of victims whom they could not help. They were refugees from the eastern Serengeti plains around Lake Lagarja, where starvation had depopulated whole districts. They had fled to their countrymen, who had barely enough to eat themselves. Swarms of vultures followed them, waiting for victims.

Farther north, Maasai, who had lived on what is now the Kenya border east of Serengeti to the Rift Valley (there were none in Serengeti), were now decimated. The surveyor G. E. Smith noted that "the Masai themselves attribute the cause of their decline . . . to the cattle plague [rinderpest], which devastated their country in 1888 . . . and to a bad attack of small-pox, which followed in the train of the famine caused by the wholesale death of their cattle."[6]

In the areas southwest of Serengeti, wildlife was plentiful before rinderpest arrived, as described by the explorers Speke and Grant in 1861. Buffalo and zebra were abundant.[7] Even in 1889, the next European explorers, H. M. Stanley and Mehmed Emin Pasha (original name: Eduard Schnitzer), found buffalo common as they passed through the southwest to circumnavigate Lake Victoria. But shortly thereafter, the disease arrived in East Africa, and buffalo and their close relatives such as wildebeest were devastated. Baumann mentions that as he walked across the Serengeti plains he saw the skeletons of wildebeest and, farther west in the savanna, those of buffalo.[8] Rinderpest then appeared every 10 to 20 years, sweeping through and reducing numbers each time, until the 1950s.

We have other similar observations. The famous hunter and game scout F. C. Selous (whose memorial now resides in the Selous Game Reserve, in southern Tanzania) noted that buffalo, together with kongoni and zebra, were found everywhere before 1890 in the Maasai lands of Tanzania and Kenya. But in 1890 he and other travelers saw that a few months after rinderpest had appeared in native cattle, buffalo numbers collapsed, and a decade later there were few left.[9] American president Theodore Roosevelt[10] and the Kenya game warden Blayney Percival[11] both noted the scarcity of buffalo in the 1900–1920 period. In August 1913, the hunter S. E. White walked through what is now the northern part of the Serengeti, and although he refers to seeing a few buffalo, they were scarce compared with present times. In the 1930s, Audrey Moore,[12] the wife of the second Serengeti game warden, Monty Moore, had to make a special journey to Klein's Camp in northern Serengeti in order to see any buffalo. Myles Turner, the assistant game warden in the 1950s and '60s, remarked that buffalo remained uncommon until the 1960s.

The ecologist John Ford describes the long-term impact of rinderpest on the ecology of the region southwest of Serengeti.[13] Wasukuma agricultural tribes suffered severe declines both from famine as a result of their cattle dying and from subsequent outbreaks of smallpox. The devastation resulted in the depopulation of large areas surrounding the ecosystem. Without cultivation and fires set by people (the latter an age-old practice that stimulates new grass growth and controls pests and unwanted thorny scrub), dense vegetation grew up, wildlife returned, and tsetse flies spread from the surrounding uninhabited Serengeti westward to the shores of Speke Gulf. The flies and the sleeping sickness they carried—fatal to both cattle and humans—made the land uninhabitable.

This situation persisted until the 1930s, when Julian Huxley, the noted environmentalist and evolutionist, was sent by the British government to assess the region for a prospective conservation area; he saw that bush clearing was taking place to reduce tsetse-fly numbers. This policy continued for several decades thereafter,[14] which allowed the eventual expansion of the human population in the 1950s; it has continued at an accelerating pace ever since.

———

When rinderpest, being foreign to Africa, first infected wild African ruminants, it caused a devastating die-off because all animals were susceptible. The few that survived had antibodies that protected them for

FIGURE 2.1. Collecting blood samples from a wildebeest carcass to look for antibodies to rinderpest and other diseases. The black-backed jackals shared the carcass with me, Gol Kopjes, February 1972. (Photo by Anna Sinclair.)

life, as with measles in humans (to which rinderpest is related). As time passed, a greater proportion of the population became immune until only the newborn animals (up to six months old) died from it. That was the situation by the mid-1950s.

When I started, in the mid-1960s, to look at disease as a possible explanation for the changes in buffalo and wildebeest numbers, I knew that blood samples had been collected by Plowright from wildebeest and buffalo in the period 1962–1969.[15] The big puzzle was that the virologists had found high levels of antibodies to rinderpest in some, but not all, adult animals. Those animals with the antibodies had at some time in the past been infected but had then recovered; the ones without the antibodies had not been exposed. But the virologists did not understand why only some adults had the antibodies—it did not seem to make sense[16] (figure 2.1).

At the laboratory in Banagi, I had the skulls of both buffalo and wildebeest from which blood had been collected over the previous six

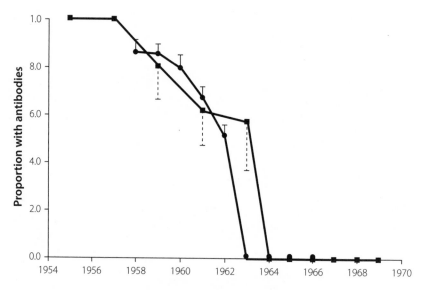

FIGURE 2.2. The prevalence of rinderpest antibodies in buffalo (squares) and wildebeest (circles) blood sera relative to the year when individuals were born. Animals born before 1963 had antibodies, while those born after 1964 did not, an indication that there was no rinderpest in the populations from that time on. (Data from Sinclair 1977 and unpublished.)

years. I was able to tell what age they were when they died by counting the layers of dentine and cementum in their teeth—each year a new layer is laid down. This allowed me to calculate in which year each animal had been born. A quite unexpected pattern appeared: in both species the animals that had the antibodies had been born before 1963 (wildebeest) or 1964 (buffalo)[17] (figure 2.2). No animals born after those dates had rinderpest antibodies and so had never been exposed to the disease; it had simply disappeared. And it was after 1964 that both species had shown the rapid population increase that scientists had noted.

The disappearance of rinderpest was at first a mystery. After being around for some 70 years, it had vanished almost overnight, catching us all by surprise. I asked Walter Plowright what had been happening over the previous decade. He told me that there had been a major international program to vaccinate cattle and eradicate rinderpest. In many areas, wildlife had been eradicated because the virologists thought that wildlife infected cattle. But they could not eradicate the Serengeti animals because of a backlash from conservationists. Serengeti, therefore, remained one of the focal points of infection, so they progressively created a ring of vaccinated cattle around Serengeti to contain the disease, and this ring was completed in the early 1960s. Immediately, rinderpest died out of the wildlife. This result was exactly the opposite of what the virologists had been expecting, namely, that cattle were protected from the wildlife by the vaccine. In fact it was the wildlife that had become protected from the cattle—*it was cattle infecting wildlife*, and not the other way around. This makes biological sense: rinderpest is a disease of cattle and over evolutionary time had developed carriers, animals that hosted the disease without showing symptoms. In contrast, the wildlife had never been confronted with the disease before 1890; they had no carriers and so either died or recovered. Once the infectious cattle carriers were removed, the disease could not stay in the wild populations, and it simply died out.

By 1970, I had calculated from mathematical models that the disappearance of rinderpest was the only likely cause that could account for the increase in wildlife numbers that had started in 1963 or 1964 (figure 2.3). This conclusion was later confirmed by the wonderfully enthusiastic epidemiologist Andy Dobson, of Princeton University.[18]

———

But what about the other possibility—that it was the legal protection of the animals, beginning in 1951 when SNP was created, that allowed the increases? If protection had been the reason for the increases, we should have seen similar increases in other species, such as zebra, which are closely associated with wildebeest. On the contrary, we found that most of the other herbivore species had not increased in number, and in particular zebra have remained at the same level of 200,000 from the 1960s to the present day (figure 2.3). Rinderpest is a disease of ruminants such as buffalo and wildebeest and does not affect nonruminants such as zebra. Thus, these trends provided the clear evidence that it was only the ruminants, which were susceptible to rinderpest, that showed the rapid increases.

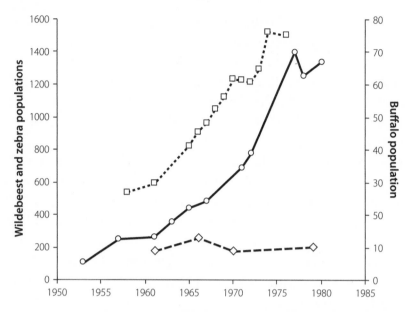

FIGURE 2.3. The increase after 1950 of wildebeest (circles), and buffalo (squares) numbers compared to those of the stable zebra population (diamonds). Numbers are in thousands.

The discovery in 1970 of the disappearance of rinderpest was the vital evidence showing that these ungulate species had been knocked down sometime in the past and were now recovering to the higher levels they presumably were at before 1890. This discovery assumed vital significance when I asked the next question: What would cause the populations to stop increasing?

3

Finding Regulation

As the evidence accumulated that rinderpest had been responsible for knocking down the wildebeest and buffalo numbers way back in 1890, and had through repeated epidemics kept numbers low until it was removed by virologists in the early 1960s, the obvious next question became: What will stop the numbers from increasing and cause them to level off? In the 1960s this remained one of the great questions in ecology, having been argued about for nearly 40 years.

To answer it, I first had to count, with the help of other scientists, the numbers of buffalo and wildebeest and determine how they were changing.[1] Counting buffalo is not as easy as it may sound. They live in herds of up to 2,000 closely packed animals, impossible to count on the ground from a vehicle—all you see is a wall of heads looking at you. The only way is to get up in the air and photograph the herds from an airplane. But the Serengeti is a large area (some 25,000 square kilometers, nearly the size of Wales), and we had to search it systematically to make sure we covered the whole area. And we needed to count the same area to make sure we could compare the numbers from year to year. We divided Serengeti into 16 blocks, each small enough to be covered by one plane in a day. We invited park wardens with their planes from other parks in Tanzania, Kenya, and even Uganda for the four-day count each year in May. It was a vast logistical effort involving fuel dumps in different areas, servicing of planes, and accommodations for the pilots. Each plane, apart from the pilot, also had an observer-photographer whose job was to plot the course on a map so that the pilot could see where to fly the transects. These were 3 kilometers apart because at 300 meters up, one can see herds out to 1.5 kilometers on each side. When a herd was sighted, the pilot flew by and the observer took a series of photos to cover the herd (if it was a large one). For good measure we also counted elephant and rhino. Careful notes on the map were made of the location of herds and the number of photos. All this was handed in at the end of the day to me as the coordinator, and by the end of the count there was a large pile of film cassettes and papers to be sorted out. Films were taken to Nairobi for printing. Then the work began; it took about two months to painstakingly sort, mark, and count the herds on the prints, making sure there were no overlaps. By three months, we had a number. This was repeated year after year for a decade and then every three or four years up to the present. We now have a 50-year record of numbers.[2]

In the photographs, the different sexes, newborns, yearlings, and adults could all be distinguished (figure 3.1). By counting year after year, we could see the trends in numbers of newborn, yearlings, and adults, and because we knew the total number from the aerial counts, we could calculate over the year the percentage of the population comprising births and the percentage of each class that had died.

Counting buffalo was laborious but straightforward. Counting wildebeest was a different matter altogether. The first scientists to attempt this, Bernhard and Michael Grzimek, in 1958, flew over the herds and guessed at the number they saw. This provided a ballpark number but was clearly insufficient for year-to-year comparisons. In 1961, Don Stewart, from Kenya, did a count by photographing the whole herd when it was on the plains in the middle of the wet season.[3] The herd was so large that it had to be covered by some 20 transects overlapping each other. Taking overlapping photos along one transect, then overlapping again on a return transect, understandably was a nightmare to untangle. This approach was repeated several times up to 1969, when we decided it was unmanageable, especially because the herd was increasing rapidly.

Then Mike Norton-Griffiths came along. A large, loud, blustery extrovert who enjoyed the good things in life, he trained at Oxford for his PhD in the behavior of bird feeding. In other words, he had not the

FIGURE 3.1. Aerial view of a buffalo herd showing the different sizes and shapes of horns used to count juveniles and females, April 1968.

slightest experience in wildlife ecology, let alone counting animals—which was a godsend. Unimpressed with the laborious, dinosaurian methods so far used, he realized that with problems this big a sampling strategy was necessary. He set about developing a sampling technique with the help of George Jolly, a statistician from Scotland, and by 1971 we had a new method. This involved attaching a downward-facing camera to the plane. Having located the boundaries of the herd from a preliminary reconnaissance, we then flew transects across the herd (we waited until the wildebeest were in a single herd on the plains, usually in April), each transect 5 or 10 kilometers apart. Along each transect vertical photos were taken at intervals of about 30 seconds. Thus, Mike had a double sampling technique—transects sampled the herd, and the photos sampled the transects. The technique had the advantage that there were no overlaps, the animals were easy to see, and there were many fewer photos to count. With a known height above the ground, using a new type of radar altimeter, we could calculate the area on the ground covered by each photo. With all these measures, we could multiply up and estimate the total number of animals in the entire herd. Sampling had the added advantage that we could put statistical bounds on the estimates—in other words, we could see how sure we were of the result. This method has been used for decades and has recently been adapted by Grant Hopcraft, who has taken over my biodiversity program and begun to explore the use of pattern-recognition software to count animals in photos.[4]

As with buffalo, we needed to know what proportion of the population comprised newborns, calves, yearlings, and adults. Instead of doing this from photos, I could count these different classes directly from a vehicle because wildebeest are often spread out and easy to see. But a huge sample had to be obtained by driving transects, preferably through the whole herd, and counting samples of up to 5,000 animals. I would drive through the herd, stopping every kilometer and counting about 100 animals each time. With the proportions of these classes and the known population size, we could, as with buffalo, calculate the proportional changes in each class from year to year. We could then answer questions like what percentage of calves and adults died each year.

———

Before I talk about our results, I need to explain a fundamental concept of populations called *regulation*. If 2 elephants produce 2 babies in 10 years, and they and their offspring continue to do so unimpeded for 100 years, there would be 1,000 of them. After 200 years, there would be 1 million, and after 300 years some 107 million. Clearly, even the slowest-reproducing animals have enormous potential to increase in numbers; and if elephants can do this, think of what insects can do. Yet the world is not overrun by elephants or any other species. So this raises the question: What limits the numbers of animals, and also plants for that matter? This was the problem that the cleric Thomas Malthus addressed in 1798.[5]

Finding the answer to this question produced the most profound debate in population ecology of the first half of the twentieth century.

Since the early 1900s, scientists have been asking what causes animal populations to change in number and what prevents population extinction.[6] First attempts to deal with the problem focused on insects because these produced population outbreaks that damaged agricultural crops, especially in Australia. In the 1930s and 1940s, Australian entomologists, particularly H. G. Andrewartha and L. C. Birch, working on tiny, millimeter-long insects called thrips, which live in flowers, noticed that when environmental conditions, especially the weather, were good survival was also good and so numbers increased. The reverse held when conditions were bad. They argued that high numbers were seen but that they always came down again; yet populations generally did not go extinct because the environment fortuitously changed in time to allow an increase again. This idea was contested by other scientists on the grounds that changes in the environment depended on random events of weather alone, and mathematicians had shown that random events led to extinction.[7] The weather idea was rescued by arguing that a local extinction was reversed by animals coming in from nearby populations. However, this immigration merely slowed down the inevitable trend to extinction.

Because of this theoretical difficulty, a different idea was proposed by another Australian, A. J. Nicholson. In an iconic paper in 1933[8]—perhaps the most important on this subject produced in the first half of the twentieth century—Nicholson proposed that populations were limited by resources such as food or space; as a population increased, there were fewer resources per individual, and eventually individuals either stopped reproducing or they died. Either way, the population would stop increasing. The argument also worked in reverse: declining populations would experience increased food per individual, which would increase reproduction and survival, and extinction would be avoided. Nicholson explained that it was the *proportional* mortality or reproduction that changed with population size; in other words, an increasing *percentage* of a population died as it increased. This percentage was directly related to or dependent on population size, and so he defined this mortality (or reproduction) as *density dependent* (i.e., dependent on the size of the population), and the populations experiencing this proportional effect as *regulated*. The regulation hypothesis got around the difficulty of extinction and explained how populations could exist for much longer than would be predicted by random events.[9]

The two schools of thought argued for 40 years, the "weather school" (Andrewartha and Birch) continuing its work with insects in Australia and Nicholson's "regulation school" broadening to include bird studies in England under David Lack.[10] In the 1950s and '60s, a twist on the regulation idea emerged from studies of animal behavior in rodents in America, England, and Canada, and of birds in Scotland (the "social-behavior school").[11] These scientists proposed that animals could evolve behavior that would prevent extinction of the group (or population). These behaviors, which also included physiological mechanisms, would kick in when populations became too large so that animals would stop reproducing in time to prevent starvation. This was regulation via behavior rather than by the direct effects of a lack of food.

By the mid-1960s, there were plenty of theories but a dearth of evidence from real cases in nature, which of course was required to resolve the arguments. The fundamental reason for the lack of evidence was that the causes of population change are difficult to pin down. They require a detailed knowledge of both births and deaths, and the factors that influence these such as the amount of available food and what kills the animals. In essence, measures of proportional mortality would show whether it was density dependent (predicted by the regulation school) or not (predicted by the weather school). If regulation was caused by mortality from starvation, this would support Nicholson's resource idea, but if regulation was caused by a lack of reproduction due to social and physiological factors, this would support the social-behavior school. So it all boiled down to measuring mortality and analyzing its causes, and to measuring changes in reproduction.

Regulation, therefore, was the key to understanding how populations persisted. But demonstrating this process required a particular situation. The population had to be changing, preferably increasing after having been reduced by some external disturbance such as a drought, an exotic disease, or maybe human hunting. If the population had been increasing because the environment was gradually improving and thus providing more food, then the food per individual would remain the same and we would not have the conditions for detecting density dependence. As a result, it was necessary to establish beforehand that the population was below the level that could be sustained by its food supply.

And this is where the discovery of rinderpest became so vitally important. Because rinderpest had knocked down the populations of buffalo and wildebeest to very low levels for many decades, I knew that they were well below their food limit. When rinderpest disappeared, they were able to increase again, which allowed me to see whether the percentage changes in calves and adults were density dependent or not. Did the proportion of calves or adults dying increase as the population increased, thus allowing it to level off? Or were they simply at the mercy of the weather? Because of the disappearance of rinderpest, I was now in a position to test one of the great ideas in ecology.

———

The counts of buffalo started at about 15,000 animals in 1961, reached a peak of 76,000 in 1975, and showed signs of leveling out thereafter (figure 2.3). At the same time, I was measuring the percentage of the population being added to by births and reduced by deaths of adults and juveniles. It took many years to get enough measurements of these births and deaths to see what was going on. But after eight years it was clear that a greater proportion of adult buffalo were dying as the population increased, which meant that adult mortality eventually would equal the birth rate and the population would stop increasing, as indeed it already appeared to be doing in 1975. In other words, adult mortality was density dependent, and therefore the population was regulated. The birth rate, however, did not seem to change; it remained at about 75 percent pregnancy each year, which was the maximum to be expected in this species, indicating that regulation via behavior was not operating. Yearling mortality was highly variable and quite unrelated to population size. In fact, it was much more related to the vicissitudes of the weather—wetter-than-normal dry seasons meant good yearling survival, and vice versa.[12]

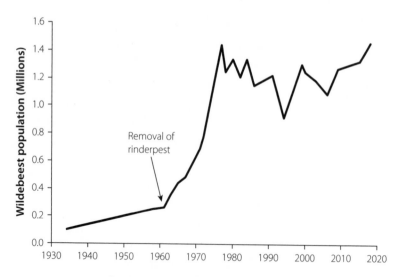

FIGURE 3.2. After the removal of rinderpest, the wildebeest population
increased rapidly from the 1960s until 1977, when numbers leveled out at about 1.3
million. The population has remained at that level during the subsequent 40 years.
The drop in numbers in 1994 was due to the great drought.
(Data from Sinclair 2012 and unpublished.)

Wildebeest took even longer for us to understand. The first reasonable counts, in 1961, gave a figure of 250,000 animals. By 1967 this had increased to 400,000 animals, and then nearly doubled to 750,000 by 1973, when we did our first double-sampling aerial photography with Mike Norton-Griffiths. Four years later, we were stunned to count 1.4 million—I sent Mike a telegram when I got the result because it was a world record for an ungulate population. In the space of 17 years, the numbers had increased almost sixfold (or 600 percent).[13] Mike and I were so surprised that we repeated the count the next year, but got almost the same result, and then two years after that, in 1980, also the same number. So this was the second great surprise: the population had leveled off, just like that, quite suddenly. And in fact the population has stayed at about 1.3 million for more than 40 years—unequivocally stable (figure 3.2). In later years, Markus Borner, of the Frankfurt Zoological Society, took over the flying duties from Mike, and continued to do so until he retired, and Grant Hopcraft took over from me the task of calculating the population size.[14]

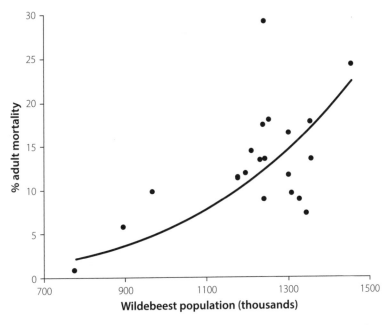

FIGURE 3.3. The percentage of the population of wildebeest dying increased with population size, causing the population to level out. (Data from Mduma, et al. 1999.)

Because I had photos from the counts starting in 1962, I was able to calculate the percentage rates for births and deaths, as I had done for buffalo. But it took me 15 years to get my first results, in 1977, showing that the percentage of adults dying was increasing as the population numbers increased. And it took another 20 years to complete the story, in 1997. This was put together by my PhD student Simon Mduma, who joined me in 1988 first as a student and then as principal investigator. Ultimately Simon became Director-General of Wildlife for the Tanzanian government. He is still with our biodiversity program.[15] With almost 35 years of data by 1999, Simon could show definitively that the percentage of the population dying as adults was, like the rate for buffalo, increasing as the population increased. Thus adult mortality was strongly density dependent (figure 3.3). On the other hand, yearling mortality was extremely variable, while pregnancy rates remained very high (90 percent) and invariable. This was the same result as that for buffalo some 20 years earlier.[16] And, if nothing else, these results showed that it takes a very long time to see how ecosystems work.

In 1977, just as our count reached 1.4 million, my colleague Ray Hilborn (remarkably, from the University of Washington College of Fisheries)[17] modeled the population using the numbers I had provided and predicted that the wildebeest would level out at 1.5 million[18] due to these density-dependent processes derived from food supply. At the time, I was skeptical because the last figure we had available was only 750,000, showing that the population was increasing at a fast rate, but Ray predicted this was where it would stop, and it did. Given the errors in sampling and counting, Ray's prediction was a remarkably close fit to what later transpired. Ray was characteristically modest and said it was just luck—but of course it wasn't. The data were telling us that regulation was happening.

———

What is the significance of these density-dependent adult mortalities? First, with a constant birth rate the increasing adult mortality rate will eventually equal the births and the previously increasing population will level off; it reaches what we call an *equilibrium*. We saw exactly this in the Serengeti wildebeest population, which increased from 250,000 in 1961 to 1.4 million in 1977 and then leveled off at about 1.3 million for the next 40 years (figure 3.2). Unfortunately, buffalo were never allowed to remain at their equilibrium because they were once again knocked down by a 20-year period of near-catastrophic poaching in the 1970s and '80s.[19]

Second, regulation allows the persistence of populations far longer than would be expected from purely random events.[20] This persistence arises because one life stage (in this case, adult survival in wildebeest and buffalo) can counteract disturbances from the environment by improving survival when numbers are low and reducing survival when numbers are very high; this keeps the population within bounds. These bounds are set by the mean level of the environment—if overall an area has lots of food, then average densities of buffalo are high and the boundaries are correspondingly high, and vice versa. Thus, in Africa we see a correlation between average rainfall (which determines food supply) and average density of buffalo. This correlation is in itself evidence of regulation taking place.[21]

There are other demonstrations of regulation—we have seen it in lion populations of Serengeti; in elephant populations of Kruger National Park, South Africa; in deer populations of Scotland; in birds in England and the Netherlands; and even in the thrips populations of Australia studied by Andrewartha in the 1940s.[22]

It has not yet been possible to obtain evidence of regulation in all the other species in Serengeti because of the amount of data and funds necessary, but if one sees regulation in wildebeest, buffalo, lions, elephants, birds, and even insects, one must expect it to be operating in all the other species and ecosystems around the world. Regulation, then, is the fundamental process underlying the persistence of all our ecosystems.

4

The Discovery of Food Regulation

The story of how I came to identify food regulation begins in November 1966, when Myles Turner, warden in charge of antipoaching in Serengeti, suggested a safari to the north of the park. The short rains had failed and the wildebeest were still in the Mara Reserve and northern Serengeti, providing great opportunities for the small groups of poachers snaring animals to take meat back to their villages. Myles wanted to catch them, and he suggested a trip to learn about the area. We spent a week there going out searching every day. It was quite noticeable that there were numerous dried-out carcasses of wildebeest and zebra lying around. These had not been devoured by predators (as I will explain later) and remained like mummies on the ground. Myles mentioned that this happened every dry season. It was clear that we were looking at mortality; the animals were just waiting for us to count and autopsy—not these, because the remains were too old, but if I could get to them sooner, just as they died, then I might be able to see what was killing them.

I had to find out what was killing animals because I needed to know what the underlying cause of the regulating mortality was. Of course, in 1966 I had yet to discover regulation, but I still had to know what was killing animals. As described in the previous chapter, much of the argument around regulation could not be resolved because there was so little evidence on the cause of death from the studies of small animals. With the carcasses of these large animals now available for examination, it seemed that the causes of mortality were for the first time within our grasp, and so we planned to return in the next dry season, August 1967, to count animals and find out what they were dying from.

First, I had to learn what clues to look for in the carcass to determine what had killed the animal. The obvious possibilities were predation, starvation, and disease. It was extremely rare that I found an intact carcass, one that had not been torn apart by predators and vultures. But when this did happen, I could take blood samples and make slides for later microscope examination of blood parasites. Far more frequently, an animal dying in the night would be nothing more than a skeleton by daybreak, with perhaps some skin remaining around the head and neck. In these cases I was able to distinguish predation from scavenging by careful detective work of the site. Animals caught by predators leave large pools of blood on the ground, while those that die from other causes leave no blood. Predators always remove the stomach and pull it aside; they don't like the large wad of fermenting grass. The stomach remains in the carcass in animals dying from starvation or disease. Lions leave claw and bite marks on the head and neck; hyenas crack open bones (lions never do).

So it was possible by these clues to distinguish predation from nonpredation. But then the problem was how to tell whether lack of food was the cause of death. Initially this proved difficult to establish because it seemed no one had ever considered the issue. There was a hidden bias among wildlife workers that with all that grass waving about at the end of the rains, there could not possibly be a shortage of food—predation must explain everything. To test this, what we needed was something in a carcass that told us whether the animal had been starving or not. At the beginning, in 1966–67, a search of the literature turned up nothing. Then I caught a break from a most improbable source. In an obscure issue of the *New York State Conservationist* that I found in the Nairobi Museum, of all places, was an article by E. L. Cheatum from 1949, mentioning that bone marrow was an index of starvation in deer that had died in winter.[1] Could this be the clue I needed? I asked the animal-husbandry experts at the East African Agricultural and Forestry Organization whether cows died with no bone-marrow fat. The answer was always the same: most had never looked, or they thought it improbable on the grounds that what prevailed in North American winters had nothing to do with the tropics. However, it seemed to me an eminently reasonable phenomenon and one worth following up despite the complete lack of evidence and interest.

————

Northern Serengeti, the area north of the Grumeti River at Klein's Camp, and including the Mara Reserve, had the highest numbers of buffalo and was the dry-season refuge for the wildebeest: it was where the migrants ended up at the driest time, when food was in shortest supply. This was likely to be the best place to look for mortality. As the dry season advanced in August of 1967, when the wildebeest had arrived from the south, I conducted, with the help of my wife, Anna, searches of the area for recently dead animals. This proved to be surprisingly easy, for all we had to do was be out there on the ridges at eight o'clock in the morning, when vultures started to soar—vultures did the searching for us, and when they saw a carcass they folded their wings and dropped like a stone. This behavior was so obvious that we soon became adept at spotting dropping vultures several kilometers away. The problem was that hyenas and lions (happy to get a freebie meal) were doing the same thing. So there was a race to get to the carcass before the hyenas went off with the limb bones. It was all very exhilarating and many of our colleagues were happy to join in and search with us, although they were less keen on the autopsy bit, especially if the carcass had been rotting for a few days—that part they left to us. Many a time lions, if there were enough of them, raced in and simply snatched the carcass away from us while we dove into the vehicle—well, discretion reasoned that there was always another carcass—but because of this we tried to have someone, a field assistant, keeping watch. By far the greatest number of carcasses were wildebeest, with a few buffalo, giraffe, and smaller antelopes.

I soon saw that there were different sorts of bone marrow. Some were solid, white, and very fatty, others were gelatinous, white, and opaque, and yet others were completely translucent like gelatin. What did it all mean? I collected samples of the marrow in vials and later, back in the laboratory, dried and weighed the specimens to see how much water had been lost. Then the dried marrow was placed in a glass apparatus called a Soxhlet, which is used to extract fat. After many hundreds of samples, I had a picture: the solid marrow contained almost 98 percent fat with very little water. The opaque samples were variable, containing 20–80 percent fat. The translucent samples had no fat at all, barely measurable, but lots of water. I could see that water replaced fat as the fat was used up, which meant that I could dispense with the laborious Soxhlet extraction and simply calculate the water content by drying. In the long run I realized that it was only the last category, the translucent gelatinous material, that was the critical evidence. This really speeded up the autopsies because I only needed to crack open a long bone and describe its consistency.

———

Once I had collected samples in the field and calibrated the method, I had to confirm that gelatinous marrow was due to starvation. Here Patrick Duncan joined the team in 1970. A young PhD student, boisterous, enthusiastic about everything, and with an infectious laugh, he went on to join the French government service as a scientist and by the 2000s had become one of the top wildlife ecologists in France with his own students in other parts of Africa. But in the 1970s he was studying topi in the western corridor and, like me, needed to unravel the mysteries of starvation. Together we took advantage of a veterinary program in the early 1970s that was collecting kongoni, wildebeest, and other species to look at infections of sleeping sickness (trypanosomiasis) and rabies. Patrick went out with them and autopsied the animals to measure the fat in the bodies and the bones. Putting all this together, we saw a trend in fat condition. Healthy animals had large amounts of fat around the kidneys, mesentery (the tissue that holds the guts together and a fat store), and heart, as well as in the long bones. As animals found less food to eat, they lost body fat first from the heart and mesentery, then the kidneys, and only when all of this was gone did the bone marrow get used, with the progression we had already found of solid to gelatinous to translucent. The most important part of this story was that bone-marrow fat was used only after all other was gone; it was indeed an indicator of starvation.[2]

The next thing I had to do was look at how fat stores changed in live animals during the different seasons of the year, and compare this with what I had found in dead animals. Again, by taking advantage of live animals that were being sampled for other reasons, I was able to measure the fat stores in the different sexes and ages of wildebeest. There were major changes in adults of both sexes through the year. In young adult females or those that had missed a pregnancy, fat stores were good and marrow was solid in the wet season (December–June), declining only slightly in the dry season. Pregnant females put on considerable extra fat in the last stages of pregnancy (November–February) but used this up during lactation, which lasted through the dry season, when their marrow fat declined to the intermediate stage of opaque-gelatinous. Adult territorial males put on massive amounts of fat, almost 10 times that of females in the wet season, but used it completely during the monthlong rut from May to June. By the time they entered the dry season, they were already in bad shape, with opaque-gelatinous marrow, far worse than the females. Young animals, calves and yearlings, had little fat in the first place because in growing animals marrow is used for other purposes such as making blood; thus they too entered the dry season with little reserves. By the end of the dry season, in October, all groups except nonpregnant females had little fat reserve left. But the most significant result was that I never found live animals with translucent marrow[3] (figure 4.1).

The marrow fat of animals I found dead also changed with the season. The carcasses I found in the wet season were generally in good shape in terms of marrow, most of it being solid, and these did not differ from the live animals. However, starting in July marrow fat dropped precipitously, to almost zero; nearly all the carcasses had translucent marrow, well below anything recorded for live animals. The picture was clear: if your marrow dropped to translucent, you died of starvation.

I shall deal with the role of predators in this story later, but I mention here that wildebeest found dead in the wet season with good marrow

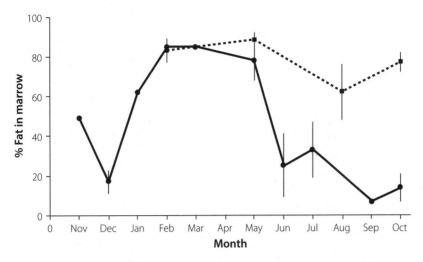

FIGURE 4.1. The percentage of fat in the femur bone marrow of live (squares) and dead (circles) wildebeest over the year. In the wet season (February–May), marrow contained high fat levels in both groups. In the dry season (June–October), fat remained high in live animals but collapsed in dead animals, showing that death was related to starvation. (Data from Sinclair and Arcese 1995.)

had all been killed by predators, while of those dying in the dry season with poor marrow, some had been killed by predators, but they were already on the way out—predators were simply speeding up the starvation process.

———

Armed with this understanding of the changes of fat stores and starvation, I set about examining the marrow of all ruminant species I found, most of which were wildebeest, with a few other species mixed in. Despite the numbers of live animals, it was always difficult to get enough carcasses to autopsy. However, this work continued year after year, and by 1977, a decade after the start, I was able to see that buffalo were dying of starvation, though I could not yet show that this was density dependent.[4] Collections continued for another decade with wildebeest, so that by 1985 I was able to demonstrate unequivocally, from 20 years of data on both the amount of food available in the dry season (figure 4.2)

FIGURE 4.2. Measuring grass food of wildebeest inside wire cages during the dry season, northern Serengeti, September 1969. (Photo by Anna Sinclair.)

and the number of animals needing the food, that starvation in adults was density dependent.[5]

Simon Mduma, whom we met in chapter 3, took over in the 1990s. Simon had the incredible good fortune of experiencing perhaps the worst drought in 100 years—the rains failed for a full year in 1993—so that by December 1993 vast numbers of wildebeest were dying. I estimated that the population had lost some 400,000 animals (there were still perhaps 900,000 left). Dead animals were everywhere, predators and vultures were satiated, and carcasses were lying about untouched. For us scientists looking to understand what killed animals, this was a gift from heaven if only we could make use of it.

Ray Hilborn, of the University of Washington, who had worked with us for decades modeling the ecosystem, was on a sabbatical with his family in our house at the time. Another larger-than-life extrovert, who liked good food and wine and went to great lengths to get them in this

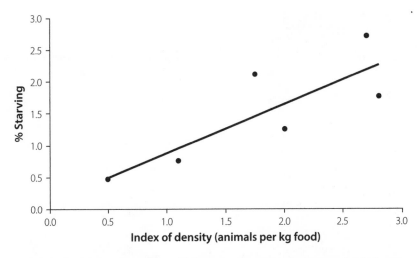

FIGURE 4.3. The percentage of the wildebeest population starving increased with density (animals per kg of food), which resulted in regulation. (Data from Sinclair, et al. 1985.)

remote place, and who got up to all sorts of escapades, Ray, together with his wife, Ulrike, and daughter Anne, and my wife, Anna, and daughter Catherine, were all dragooned (though not without protest) into searching for and cutting open the sometimes-rotting carcasses to help Simon—we had many vehicles in the field that season. It was a bonanza of data that made Simon's thesis. He finally synthesized some 33 years of bone-marrow data showing not only the density-dependent starvation, but also that the food eaten by wildebeest in the dry season was less than what they needed to survive.[6] All this was possible only because the wildebeest population (and that of buffalo too) was increasing, regaining its equilibrium from the massive disturbance of rinderpest decades earlier. This saga, starting in 1966, finally came to a conclusion in 1999, and all thanks to the clue provided by Mr. Cheatum half a century earlier. This work showed for the first time that adult mortality due to starvation was regulating the wildebeest population (figure 4.3).

───────

The story does not end there. It gets even more interesting because the food-regulation process also involves predators. They do kill buffalo and

FIGURE 4.4. Starving buffalo were easily killed by lions during the great drought, December 1993.

wildebeest, especially those weak from starvation, but predation accounts for only about a quarter of the deaths, the rest coming from starvation (figure 4.4). However, predators influence how this starvation occurs. Our long time series of data on wildebeest bone marrow covered both the increase phase (1966–1977) and the stationary phase (1977–1999) of the population. During the increase there was plenty of food and little starvation to begin with, but starvation gradually increased as numbers increased. In the stationary phase, populations were at their highest and consequently starvation was at its most extreme. The subtle part of this story is that the environment containing the food, the grasslands in this case, is not uniformly accessible: some parts are safe from predators, while others are dangerously exposed—open grassland in the former, and areas close to rivers and thickets in the latter. Obviously animals prefer to stay in the safe habitat if they can.

Here we meet again Grant Hopcraft, who helped us when we were counting buffalo and wildebeest; Grant is one of the most accomplished field men in Africa I have seen. He started with me as a student in 1997 at the University of British Columbia, was then at Groningen University in the Netherlands, and finally joined the University of Glasgow as a researcher and faculty. He recorded sites where lions killed their prey, using the huge data set of Craig Packer, from the latter's long-term studies of lions,[7] and found that lions were most successful near thickets from which they could ambush prey—although these areas had the fewest prey because prey would avoid them if they could. Lions did not usually go to where the most animals were, out on the open plains, because there was nowhere from which to ambush.[8]

Thus open grassland is safe and areas near thickets are not. But there is less food in safe and more food in unsafe habitat because of the difference in grazing intensity. Animals use fat stores to carry them through the dry season (July–October), when, as we have seen, there is not enough food. A healthy animal's store of bone marrow is just sufficient to carry it through to the end of October while staying in safe habitat. However, animals that enter the dry season with a low fat supply—for example, the breeding males—do not have enough to carry them through in safe habitat; their bodies seem to be able to compute this at the beginning of the dry season, and so they know to move early to (necessarily) unsafe habitat where they can get more food and slow down the rate of fat use. Naturally, some of these are killed by predators. We see the consequences of this predator-avoidance behavior when we look at the bone marrow: those killed early in the dry season have better marrow fat than those killed at the end, as might be expected. But here comes the twist: during the population-increase phase, those killed were taken later in the season and had moderate to poor marrow, because with all the good food they could wait longer before moving to unsafe habitat. In contrast, during the stationary phase there was so little food that animals had to move out much earlier, when their marrow was still relatively good, so the average marrow of predator kills over the dry season in the stationary phase was *higher* than the average marrow level of those killed in the increase

phase. It is counterintuitive, but that is how predators can alter the behavior of prey to determine who gets killed.[9]

————

Scientists recognize that there are two major ways that regulation can
occur. The first is through the flow of energy and nutrients from the soils
through the plants to the herbivores and on up to the predators. This is
called *bottom-up regulation* and operates through *competition* between
individuals for limited supplies. The second is through predators eating
herbivores, which indirectly affects plants and even soils. This is called
top-down regulation; it operates through *predation*, and produces what
is called a *trophic cascade*, an ecological term coined in the 1980s to describe reciprocal changes in food webs (*trophic*, from the Greek, refers
to nourishment). This refers to the cascading effect that predators have
on prey on the level below them, which in turn affects the plant level.
These levels, made up of particular species, energy, and nutrients, are
called *trophic levels*.

Competition can only occur if food or other resources are insufficient
for the population. At low population numbers there should be plenty
of food for all, but as the population increases there will be less and less
food per individual if the food supply does not increase also. Eventually
an increasing proportion of the population dies from starvation so that
it becomes *food regulated*. The regulation of buffalo and wildebeest,
therefore, was bottom-up—food determined the size of the populations, through competition between individuals.

————

While I was counting buffalo each year, I also counted elephant and
found that that population was also increasing rapidly, from a few hundred in 1958 to 3,000 in 1973. At first I did not understand the cause of
the increase, but after I became aware of the early history of the ivory
trade, I understood why. In the early 1800s there was a development in
the ivory trade that was to change the face of Africa (including Seren-

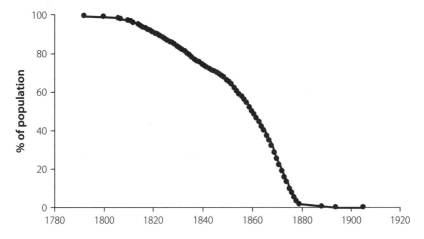

FIGURE 4.5. Although there have been ivory exports from East Africa since Roman times, the amounts were low and stable, so we can assume a high, stable elephant population in the 1700s. Exports of ivory increased rapidly from 1840 to 1880 and then collapsed completely when elephant populations were wiped out, as the cumulative curve of remaining live elephants shows. (Data from Sheriff 1987.)

geti) by the end of the century. Arab traders had for centuries been exporting ivory and slaves to Arabia and India, but on a relatively small scale. This began to change when, in the mid-nineteenth century, the sultan of Oman moved to Zanzibar Island and made it his permanent home. It was from that time that trade with the interior of East Africa expanded rapidly. Both ivory and slave exports rose exponentially.[10] The hunting was so heavy that elephants were eradicated from most of Africa[11] (figure 4.5). This extirpation started near the east coast in the 1850s, then extended westward as the caravans were obliged to travel farther to find ivory.[12]

Recovery of the elephant population in East Africa took the first half of the twentieth century, starting from the few remaining groups in remote areas. Tribal elders remembered seeing two elephants in the west of Serengeti in the 1930s, the first they had ever seen. It was not until the 1950s that elephants were commonly seen both in and out of protected areas. Subsequent counts showed that groups were invading Serengeti both from the northern forests of Kenya and from the southern (then uninhabited) savanna south of Maswa. And recent DNA analysis shows that these two groups had different genotypes. In the early 1950s, Serengeti had fewer than 100 elephants, but by 1958 some 600 had immigrated and in 1961 counts recorded over 1,000. By 1965, elephants had spread throughout the Serengeti ecosystem and numbers had increased to 2,000. A decade later they had reached 3,000.[13] The dense vegetation that was a feature of savanna Africa in the first half of the twentieth century provided abundant food to support this immigrating population, allowing a high birth rate (as measured from photographs used for counts), and the result was a dramatic increase in numbers. The same increases were observed in most savanna areas of Africa where elephants lived. Although some of this increase must have been due to compression from expanding human activity outside, it was certainly not the only cause of the increase, or even the main cause; the populations were growing back from the low levels of a century earlier.

———

So elephants had been cleared out of Serengeti, and I realized that they too had been reduced below their food limit by ivory hunting and then, with protection (after the park was formed in 1951), began increasing back to a level determined by food. They were then knocked down again by massive poaching in the 1970s and '80s, to only about 300 animals by 1986—a 90 percent loss (though some moved into Kenya, where poaching was less widespread). Once again, the population increased after the trade in ivory was banned in 1988 (figure 4.6).[14] In both cases of recovery from hunting, the rate at which they increased gradually lessened, indicating that they too were regulated, a feature seen also in Kruger National Park, South Africa, and Hwange National Park, Zimbabwe.[15]

Regulation of elephant numbers takes place through lack of food in the dry season, mediated by the availability of water. The mechanism was seen most clearly in Tsavo National Park, Kenya. Populations there were increasing following the imposition of protection from hunting in the 1950s. Their food in the dry season consisted of shrubs and trees within reach of water supplies, because water is an essential resource. As numbers increased, they first consumed trees near water, then pro-

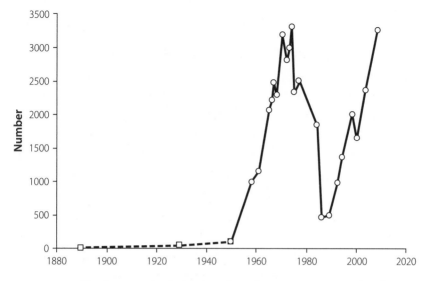

FIGURE 4.6. The changes in the elephant population from 1880 to 2008. Numbers from
1880 to 1950 are estimates; those subsequently are from aerial counts. Numbers gradually
increased after the ivory trade collapsed in 1880 and reached a peak in 1974.
There followed another episode of severe ivory poaching from 1974 to 1988. In 1988
the ban on trade in ivory resulted in a rapid increase of numbers again to present times.
(Data from Sinclair, et al. 2007 and unpublished.)

gressively farther away until they were only just able to reach their food.
In 1971 a drought resulted in insufficient food within reach of water, and
a large proportion (perhaps 30 percent) died of starvation. The spatial
distribution of elephants relative to food within reach of water supplies
was also the important feature regulating elephants in Hwange Park,
Zimbabwe. Besides elephants, there is evidence of food regulation of
white rhino in South Africa and hippo in Uganda.[16]

The discovery that large herbivores were regulated by food supply
was particularly important because in sufficiently large natural areas
(such as the large national parks), populations could level out without
human interference. This avoided the particularly contentious practice
of the culling of, for example, elephants, as has been the practice in
southern Africa (in very small reserves that act more like zoos, numbers
have to be managed just as in zoos because the ecosystem as a whole is
missing).

———

So far, I have described the discovery of bottom-up processes from plants to herbivores. But of course limitation of food can also occur at higher levels in the food chain. The availability of food for predators is also a matter of bottom-up regulation. My colleagues George Schaller, Brian Bertram, and especially Craig Packer watched lion numbers increase on the plains over the period 1966 to 2005.[17] Similarly, Hans Kruuk and others observed hyena numbers increase.[18] Both coincided with the increase in wildebeest numbers. Lion numbers are correlated with prey availability in the so-called poor season, which is when migrant populations are absent from lion territories. However, migrant wildebeest leave behind them sick and aged members, which do not migrate and so become part of the resident prey. Grant Hopcraft showed that, as wildebeest numbers increased and as they ran out of food, a greater number of stragglers became available, thus providing a bottom-up increase in food for the major predators.[19] In general, this conclusion applies elsewhere in Africa where the numbers of the major predators— lion, spotted hyena, cheetah, and leopard—were positively correlated with the populations of their main ungulate prey.

———

Bottom-up regulation of herbivores is consistently a feature of ecosystems with large mammals, and especially the very large mammals, the megaherbivores like elephant, rhino, hippo, and buffalo, as we will see (chapter 7). Having outgrown their predators, they are no longer subject to top-down regulation.

Africa may be the last remaining place where megaherbivores dominate, but we should remember that these huge species occurred throughout Europe, Asia, and northern Africa only a few thousand years ago, in both North and South America a mere 10,000 years ago, in Australia some 50,000 years ago, and in New Zealand only 500 years ago. Bottom-up regulation may have been far more prevalent than we have appreciated, and its absence now is an artifact of humans' takeover of the world.[20]

5

How Predators
Regulate Prey

By the early 1980s, it was becoming clear that bottom-up regulation of buffalo and wildebeest was being driven by the food supply. We concluded that competition for food both within a single-species population and between species populations was the process shaping the Serengeti community.[1] However, there were some outstanding inconsistencies in our observations that were not predicted by our competition conclusion. In particular, numerous other species—the smaller herbivores such as topi, kongoni, impala, and the many even smaller antelopes—clearly were living together in tight, multispecies groups, and all were eating grass to some extent. As many as six species could be found together in the same group. Their diets were not exactly the same, but there was far more overlap in diet than one would expect given that competition was shaping their food niches. Inconsistencies of this sort tell scientists that they do not have the whole story. Competition was only a part of the story.

The problem was that we knew nothing about these other antelopes. What process was regulating them? Was it predation after all? This was obviously the next big question—it was a vital gap in our knowledge of how the ecosystem worked. Large species were easy to find when they died. Their carcasses lay around and we were able to conduct autopsies. Small species, however, were dismembered and eaten fast by scavengers such as hyenas, jackals, and vultures, and nothing remained of the carcass within an hour. We knew little about the causes of their deaths.

How to study the other species became, therefore, the outstanding problem. I realized, however, that perhaps the species themselves could tell me, by their behavior, whether they ran out of food or were limited by predators. Because the wildebeest were limited by their own food supply in the dry season, any other species eating the same grass food would be competing with the wildebeest. And I knew that at least with topi and kongoni (which are closely related to wildebeest), and sometimes with impala, they did indeed eat the same food.[2] If they competed, then competition theory predicted that these other antelopes should avoid the wildebeest in the dry season when food is limiting but not in the wet season when food is plentiful. In other words, these antelopes should be found farther away from the wildebeest in the dry season than in the wet season. In contrast, if predators were regulating these species, they would be keeping antelope numbers below the limit set by food supply. Ergo, there should be no competition for food between antelopes and wildebeest, and antelopes should not move away from wildebeest in the dry season. In fact, I suggested, antelopes should gravitate *toward* wildebeest herds because of the rule of safety in numbers, and especially because I knew from the work of Brian Bertram that predators prefer to eat wildebeest over any other prey. Thus if you are a topi on your own, lions would have no choice but to eat you, but if you stand next to a wildebeest, the lions would always choose the wildebeest, not you. So best to move toward wildebeest. The important point is that the two ideas, food regulation and predator regulation, made opposite predictions about the behavior of the antelopes when the wildebeest migration appeared in the dry season in the Mara Reserve. Testing each prediction merely required a detailed record of where each species occurred and what they did in relation to the presence of wildebeest.

I tested these predictions during the dry season of 1982 by measuring how far the different antelope species were found from the wildebeest as the season progressed and the food supply grew short. And, indeed, I found that topi and kongoni actually moved closer to the wildebeest as food became scarcer. The results suggested that predators were having a far more important effect on the antelope numbers than we had previously thought. Predators could be keeping the numbers of these smaller antelope species low enough that they had sufficient food and so were not competing with wildebeest. And because there was enough food, these antelopes were staying close to wildebeest for protection.[3] This was our first real clue that predators were important in determining numbers of some antelope populations. It changed our view of what was shaping the Serengeti ecosystem. Competition was certainly taking place in large species such as buffalo, and also in migratory species such as wildebeest, but it looked like predation was shaping the rest of the community.

However, although these observations were suggestive, they did not provide definitive evidence that predation was regulating the smaller antelopes. The effect of predators on prey numbers is complicated. Yes, they eat animals and so reduce prey numbers, but their consequences can be *regulatory* (stabilizing) or the opposite, *antiregulatory* (destabilizing), which means that if predators can reduce a prey population, their effects get stronger and stronger until extinction occurs. Depending on circumstances, these effects can switch from stabilizing to destabilizing on the same prey, or can be stabilizing on one prey species but destabilizing on another species.[4]

––––––––

Together with Simon Mduma and Peter Arcese, a postdoctoral fellow who had just joined my team, we set about trying to solve the problem of whether small antelopes were regulated by predators. To do this, we needed to find out what was killing them, despite our not being able to find the carcasses. By the mid-1980s newly developed techniques were coming to our aid. In particular, radios attached to collars signaled both where an animal was and when it died; we could hear the signal on a receiver, and I could detect the direction from which it came. This allowed us to track the animal. The signal was emitted at a particular pulse rate, which doubled when the animal was stationary for a certain length of time—a time that indicated it had to be dead. So when we heard this

FIGURE 5.1. Oribi male with mortality collar used to detect predation,
northern Serengeti, October 1988.

new fast pulse, we searched for the radio. From the carcass, we could
detect whether it had died from predation, disease, or lack of food.
Sometimes only the collar remained, but other clues allowed us to tell if
the animal had been killed or simply scavenged. For example, as noted
earlier, when an animal is killed by predators it bleeds, leaving a telltale
patch of blood on the ground. If the animal had died before predators
found it, then it would not bleed on being dismembered. Cats, such as
leopards or lions, leave claw and tooth marks on the neck and head, while
hyenas do not; cats do not eat collars, while hyenas sometimes chew on
them. The state of the collar gave us a clue as to who was eating the animal. Detective work, therefore, allowed us to build up a picture through
the use of radio collars of what was killing antelopes. We focused on the
three common species of topi, impala, and oribi (figure 5.1).

The results from these records of dead animals showed quite clearly
that all of the smaller and midsize antelopes died from predation; we
did not find any with our radios that had died from starvation or other
causes. This evidence showed persuasively that predators were account-
ing for most if not all the deaths in these smaller resident antelope popu-

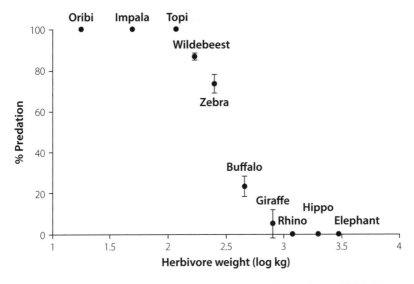

FIGURE 5.2. The percentage of deaths caused by predators relative to the size of the species. Small antelope up to the size of topi are all killed by predators, whereas adult large herbivores are almost never killed by predators and die of starvation. There is a rapid change in these causes of mortality when the average weight of a species reaches about 150 kilograms. Data for wildebeest are from nonmigratory populations. (Redrawn from Sinclair, et al. 2003.)

lations and so must be regulating them (figure 5.2). It was strong circumstantial evidence that predators were holding their populations below the level of their food supply.[5] But again, it was not conclusive proof.

———

As so often happens, a lucky (if sad) break gave me the chance to test the predation theory of regulation. In the 1980s, during a period of massive poaching because of a breakdown in protection, predators in northern Serengeti had been almost eradicated (because they had gotten in the way of other poaching activities on elephant, rhino, and buffalo). I had been counting in this area resident populations of small ungulates such as oribi (10 kilograms), Thomson's gazelle (25 kilograms), warthog (40 kilograms), impala (50 kilograms), and topi (120 kilograms) from the 1960s up to 1980, during which time predators were abundant, and then from 1980 to 1988, when predators were reduced in number due to poaching. I continued counting after 1988, when illegal hunting once

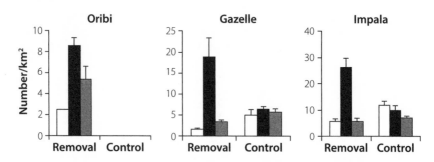

FIGURE 5.3. The consequences of predation can be seen on small antelope such as oribi, Thomson's gazelle, and impala. During the 1980s in northern Serengeti predators were killed by poachers (Removal area). During this decade numbers of all three antelope increased (black bar) relative to the period before predators were killed (1967–1980; open bar). After 1988 predators started to return, and antelope numbers declined again (gray bar). These changes in numbers were not seen in the adjacent Mara triangle of Kenya (control area), where predators were never reduced (oribi did not occur in the Mara). (Redrawn from Sinclair, et al. 2003.)

more came under park control and predator numbers built up again. At the same time, I had counted these species in the Mara Reserve of Kenya, which was immediately adjacent to northern Serengeti. Protection against poachers was much stronger in Kenya than in Tanzania at that time, so the Mara did not experience any decline in predator numbers.[6] This was the perfect test for the effect of predators because I could compare northern Serengeti herbivore numbers before, during, and after predator removal, and also compare these three time periods with numbers on the Mara side, where there was no change in predators.

I got a surprising result: for all the smaller species up to the size of kongoni (120–170 kilograms), numbers increased during predator removal and then declined again after predators returned, while in the Mara no such changes were observed (figure 5.3). This was conclusive proof that predators were determining the numbers of these resident prey. However, in the much larger giraffe, no such changes took place— they were simply not affected by predation. These results provide compelling evidence for predator regulation of small, but not large, ungulates.[7] Using the terminology I introduced in the previous chapter, this is *top-down regulation*.

FIGURE 5.4. Hyenas stealing a wild dog kill, Simba kopjes on the long-grass plains, May 1977.

Similar effects of predators have been seen in Kruger Park, South Africa, where predation was the dominant cause of mortality for small and medium-sized ungulates.[8] Other studies in southern Africa have shown that where predators have been removed on private ranches, small ungulate species have increased in number and are controlled by game-ranching programs. Similarly, in North America the removal of predators has resulted in outbreaks of deer and indirect consequences on the rest of the community.[9] In contrast, where predators have been allowed to return on ranches in northern Kenya, populations of smaller ungulates have been reduced by predation.[10]

But top-down regulation is not just about predators limiting prey numbers; they can also limit other predator numbers. Lions and hyenas are the top predators, and they kill and steal food from other, smaller predators. Cheetah cubs are often killed by these top predators; cheetah numbers, however, were not reduced by these top predators, partly because cheetahs were able to avoid them. When wild dogs were present in the system, before the 1980s, their cubs were killed by both hyena and lion, and hyena often stole their food (figure 5.4). It is now generally agreed that competition and predation by the increasing number of top predators was at least in part responsible for the extinction of wild dogs.[11] It is likely that leopards are also limited by lions; on the high plateau of the Crater Highlands, above the Ngorongoro caldera, there has been an outbreak of leopards since about 2010 because Maasai have killed most of the lions.[12]

———

Once I understood how predation caused top-down regulation in one part of the ecosystem (predators limiting small resident ungulate prey), it then became obvious that there were top-down impacts in many other components of the Serengeti ecosystem, such as the effects that herbivores have on their plant prey. For example, elephants (which have no effective natural predators themselves) have major impacts on tree populations. Perhaps the most striking illustration of this concerns the yellow fever acacia, a favorite food of elephants. These magnificent, tall (to 12 meters) trees grow in the damp grasslands near rivers. Mature trees cannot be knocked down by elephants, but when the trees are young and growing, with trunks, say, up to 40 centimeters in diameter, elephants readily push them over to eat the leaves and small branches. Until about 2012 (after which most large trees had died of old age), there were stands of these huge trees along the Seronera and other rivers. By measuring specimens of known age, we were able to estimate the age of the largest trees; we found that they had started growth in the period 1850 to about 1920. It happens that this was also the period when elephants were absent from Serengeti because of the ivory trade, which devastated the elephants of Africa in the nineteenth century (as described in chapter 4). Then followed a long period when there were no young trees growing, the same period when elephant numbers were increasing, from the 1950s to the 1970s. It seemed that yellow fever trees could grow only if elephants were not around. This idea has been strengthened by a second period of elephant killing for ivory, from 1978 to about 1993, when numbers dropped to only 20 percent of what they had been previously. In that time window, we saw dense thickets of young, regenerating yellow fever trees along the rivers. Then, for a second time, after the international ban on the trade of ivory was instituted in 1988, elephant numbers increased and are now (2021) back at their original high point—and all the young trees were eliminated, every one of them, by about 2005. So elephants are imposing a top-down limit on yellow fever trees, which are a minor component of the elephant's diet (we call these *secondary prey*).[13] Even though they do not depend on fever trees, elephants can and have caused their elimination in many river systems of the park. Instead, elephants depend on the vast numbers of other acacias, like the stink-bark acacia, which they can reduce and limit, but not eliminate—they are the main prey of elephant (and we call these *primary prey*).[14] We will revisit this story in chapter 10, but as an afterword, how do yellow fever trees stay in the system at all? Almost certainly, they normally live in small, isolated pockets that elephants find difficult to reach, protected perhaps by steep banks, rocks, and hills. These trees only expand along the rivers when elephants have been removed for long periods.

We can also see evidence of the top-down trophic cascade of ungulate grazing on plants following the collapse of wildebeest numbers in the 1890s as a result of the rinderpest outbreak. With many fewer wildebeest, a widespread increase in the amount of grass (biomass) was observed, with two distinct consequences, depending on habitat. On the eastern Serengeti plains (currently a short-grass sward of about 10-centimeter height), there was a change in structure in the 1930s–1950s period,[15] when tall grass (50–70 centimeters high) took over. Even after the wildebeest population was reconstituted, an experimental 10 × 10-meter fence (set up on these plains for 10 years prior to 1986) generated a similar tall-grass sward (figure 5.5). Indeed, much of the research by Sam McNaughton and his team has confirmed that wildebeest determine which grass and herb species make up the grassland community.[16] The change in grass biomass on the plains from short to tall grass has several indirect cascading consequences. Among these changes was a major drop in flowering-plant (dicot) diversity from

FIGURE 5.5. A 10-year grazing enclosure on the short-grass plains at Southeast Kopjes showing the effect of wildebeest grazing outside and the change to a long-grass plant community inside (May 1986).

around 70 species on grazed patches to only 15 species in ungrazed swards, because grass, once released from grazing, can outcompete small, low-lying dicots for nutrients and light.[17]

The second consequence of the lack of grazing during the 1890–1963 period occurred in the savanna habitats. During the dry season, wildebeest grazing keeps the grass low, about 10 centimeters in height. Without grazing, this grass grows taller and denser and provides considerable fuel for widespread grass fires. Before the wildebeest increased in the 1960s, on average 80 percent of the northern Serengeti savanna burned each year,[18] whereas after wildebeest reached high numbers in 1977, the area burned dropped to below 20 percent. Although there was an initial increase in tree densities in 1890–1900 (due to the reduction of fires set by people who had abandoned the area because of rinderpest), this was short-lived. When people returned, starting around 1920, and began setting fires again, savanna tree numbers dropped considerably. Burning and competition from tall grass inhibit the regeneration of acacia tree

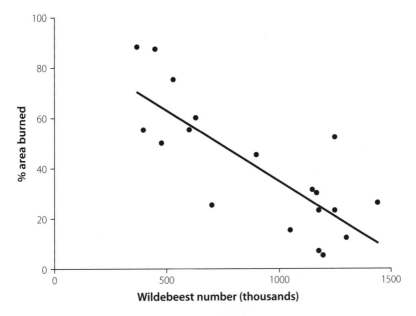

FIGURE 5.6. The area of Serengeti woodlands burned is inversely related to the size of the wildebeest population. The wildebeest ate up the grass, reducing competition with tree seedlings and leaving less fuel for fires. (Data from Sinclair, et al. 2007.)

seedlings, as Tom Morrison has found (chapter 11). When both the area burned and the competition from grass were reduced by wildebeest grazing (figure 5.6), trees resprouted freely and formed dense stands. These were evident by the 2000s: tree density in the 2000s was eight times the level it was in the 1960s (see figures 9.3A–D and figure 5.7). These acacia stands were in the 2010s being thinned out by the high elephant numbers.[19] Changes in tree populations have further, indirect consequences on birds, rodents, and insects, as is also observed in other parts of Africa.[20]

———

Elsewhere in Africa, the impact of ungulates on their plant prey has been demonstrated over many decades through the use of herbivore-exclusion fences—areas where herbivores have been kept out in order

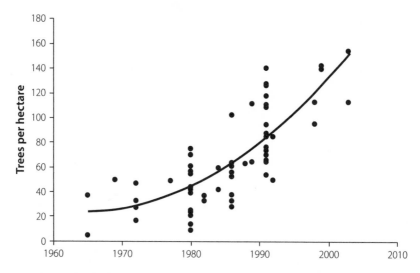

FIGURE 5.7. The number of umbrella and stink-bark acacia trees per hectare (the dominant trees) increased eightfold from 1960 to 2000 due to the reduction of area burned. (Data from this study collated with the help of Kristine Metzger.)

to measure the increased plant growth.[21] For example, in Kenya herbivores were excluded from six savanna sites for periods of five to nine years. Relative to outside the fences, there was a significant increase in tree densities, but only a marginal increase in the herb-layer cover. Associated with tree abundance was an increase in gecko and beetle numbers, as well as small mammals, suggesting strong indirect effects of ungulate predation on young trees.[22]

Top-down limitation is found in many other components of the ecosystem. One example is the impacts of birds on insects. Of the 562 most common species of birds in the Serengeti, some 104 are vertebrate feeders (eating birds, rodents, reptiles, and fish), and 343 are insect feeders, compared to 115 seed or fruit feeders (another 78 species are rare and have little impact in the system). Among birds, the food chain consists of birds of prey feeding on insectivorous birds, which then feed on insects. Ally Nkwabi, a quiet-spoken field assistant who went on to complete both his master's and PhD with us, compared bird and insect abundance on burned areas, grazed areas, and undisturbed areas. He found that on the burned and grazed areas, insect abundance declined while bird abundance went up. This suggests that the more open, disturbed areas allowed birds to find the insects more successfully, thus driving the latter's numbers down while increasing the abundance of the bird predators—birds were acting in a top-down way.[23]

———

I have talked as if there were either bottom-up or top-down processes. In fact, we can see both simultaneously, a joint bottom-up and top-down interaction, called *reciprocal trophic interactions*. Although large resident ungulates such as buffalo and elephant are bottom-up regulated, they can have top-down impacts on the plants they eat, but such impacts depend on the season in which they occur. For example, buffalo in the Serengeti are food limited in the dry season, when grasses are dormant. (Even total removal of the aboveground dry grass cannot kill the plant because it has stored its nutrients safely in the roots, and consequently there is no impact on either the community of plants or their growth.) The same argument applies with wildebeest grazing in the dry season. So there is no top-down effect on plants when they protect their growing points and storage organs in the soil during the dormant season.

A different situation occurs when herbivores feed on green plants such as shrubs, herbs, and grasses during the period of food abundance. On the short-grass plains of Serengeti, the grazing impact of wildebeest takes place during the growing season, and determines (by top-down effects) the types of plant species, their height, and growth. Grazing's effect was seen clearly when rinderpest killed off the wildebeest in the 1890s, allowing grasses to grow taller and new species to come in. Large herbivores like giraffe are limited in numbers by their food supply of shrubs and trees (bottom-up limited), but they also limit the growth of their own food supply by preventing small trees from growing into tall trees, and by altering the abundance of different tree species—a top-down effect as well.[24]

In many areas of Africa, we see that elephant, hippopotamus, and white rhino also play a pivotal role in the dual top-down impact of grazing and bottom-up limitation through plant-food supply.[25] Both white rhino and hippopotamus convert tall tussock grasslands into short-grass communities, but are limited themselves by the growth of those short grasses. Joint top-down and bottom-up effects occur when African elephants feed on trees year round and have no alternative food sources. One example occurred in Tsavo National Park, in Kenya, in 1971: increasing numbers of elephants feeding on trees resulted in many trees dying (a top-down effect); but the scarcity of trees meant that elephants starved, with some 6,000 dying (a bottom-up effect). The reduction of elephant numbers then allowed the trees to increase again.[26] Similarly, in Chobe National Park, Botswana, broad-leaved woodlands originated only during episodes of regeneration when browsing by elephants and impala was low.[27] When elephants were nearly exterminated in Africa in the mid-1800s by the ivory trade (see figure 4.5), extensive tree and shrub communities developed, especially along riverbanks. When elephants returned after the 1950s, the plant composition changed and the vegetation became more open.[28]

———

By the early 1990s, we had advanced our understanding of how the ecosystem worked. We had detected regulation. In large species and migrants, regulation operated by bottom-up limitation of the food supply. Now we had recorded top-down regulation by predators, not just on small ungulates but also on many other components of the ecosystem, including smaller predators, insects, and plants. Grant Hopcraft and Han Olff, of the University of Groningen, in the Netherlands, have shown that the direction of regulation—whether top-down or bottom-up—depends on how productive the environment is (low growth versus high growth of vegetation that acts as food) and also on the body size of the herbivores (we will look at body size again in chapter 7).[29] At first sight, both directions appear straightforward; top-down produces a trophic cascade because predators regulate herbivores, whereas bottom-up regulates herbivores through limited food. But with both processes, profound complexities shape the community of species that we see in front of us. Top-down is now considered to be the dominant process in many if not most ecosystems of the world; we have seen this in oceans, forests, northern-hemisphere woodlands, and Arctic tundra, as well as our tropical savanna.[30] As noted in chapter 4 (and see chapter 7), bottom-up limitation is very much a feature of ecosystems with large, herbivorous mammals. In the Serengeti ecosystem, both regulatory pathways are present simultaneously, and one supports the other: lion numbers are boosted by the high wildebeest population (bottom-up), which allows lions in turn to limit both the smaller predators and herbivores (top-down). This is one of the interesting features of Serengeti.

6

How Migration Structures Serengeti

While studies of buffalo and wildebeest regulation were progressing in the 1980s, one outstanding question confronted scientists: What causes migration? Serengeti, known for its lions in the 1920s, later became famous for the great migration of wildebeest and zebra. Buffalo are resident animals, wildebeest are migrants. Both are regulated by food. Why does one migrate and the other not? Why does Serengeti (and a few other places in the world) have this feature, whereas most other ecosystems do not? What is so special about wildebeest that allows these movements? What are the consequences of migration for the animals? Few species migrate, especially among mammals; what drives the ones that do? What, indeed, are the principles that produce this phenomenon? I had to find answers to these questions if we were to understand why Serengeti is unique. The wildebeest migration could truly be called one of the seven wonders of the natural world. Why?

Before we go any further, we need to understand what is meant by *migration*. Simply put, it is a seasonal round trip; animals move with the seasons to one place and then return. This excludes, therefore, mass movements such as locusts that breed in Sudan and then move south with the rainstorms, causing havoc with crops—this is a one-way trip, an emigration. Another example of this is the emigration of the red-billed quelea, a small bird (about 12 centimeters long) that feeds on grass seeds and especially crops such as millet. A single colony can contain a million nests (we have found this in western Serengeti), and after eating all the seed in an area and destroying the crops in the process, the birds move on and start again—nesting and emigration take only about three weeks. Migration also does not include the nomadic movements of grazing mammals, birds, and insects, mentioned earlier, that follow rainstorms.

———

Although Serengeti is now famous for its iconic migration of ungulate herds, the world was not even aware of it until as late as 1960, when Professor Bernhard Grzimek and his son, Michael, made the film *Serengeti Shall Not Die*. Before this, Serengeti was known only by hunters as a place to shoot magnificent, black-maned lions. Hunters and wardens alike, from the 1930s to the 1950s, were aware of the ungulate herds as they appeared on the plains in the wet season, but had no idea where they went after they left in May. In Kenya, the reverse was true: in 1913, game rangers there saw the herds appear on the Mara River in the dry season and then disappear somewhere into Tanzania, but did not know where.[1] So it was not until the Grzimeks did their first surveys by plane, in the late 1950s, that we had some rough idea where the animals went (approximate because there were no maps).[2] Then, in 1962, Murray Watson arrived as part of a United Nations–sponsored research program through the Food and Agriculture Organization. He was the first person to record the areas that the migrants used in different seasons. But the records were piecemeal, showing location but not accurate numbers. To remedy this, Mike Norton-Griffiths, whom we met counting wildebeest in chapter 3, and his associate, Linda Maddock, decided in September 1969 to set up a regular aerial survey of the ecosystem and document more precisely what was happening.[3] This was a mammoth task. Each month, a plane flew over the entire area, some 25,000 square kilometers, in straight lines 10 kilometers apart. It took two days of flying 12 hours a day. Two teams of four alternated in six-hour shifts. We did this for nearly three years. By the end of it, we had better documentation of the whole system than either before or since, an enormous wealth of data that we are still using today. It provided an example to the world of the value of scientific monitoring of ecosystems.

———

Where does the migration go? Let's start on the plains in the south (see figure 1.8). In the dry season, from July to October, the plains have no food; the grass is dry and of no value. The first rains in November produce patches of young short growth that only gazelles can feed on. With more rain there is more growth, and wildebeest appear from the woodlands on the western plains, pushing the gazelles farther east in front of them. Eventually as rain becomes general, by February, the wildebeest move on to the short-grass plains, the gazelle are pushed even farther east, as far as they can go, onto the Salai plains, and the zebra settle onto the intermediate plains with longer grass, in the Simba Kopje-Naabi hill area. The herds then move around within their own particular grassland type. As they eat up one area, they move to another every few days, sometimes in the north near Lemuta hill, sometimes in the far south, beyond Lake Lagarja. At any particular time in the migration the herds move around short distances on a daily basis—that is, they are nomadic, following the rain. Generally, they spread themselves out as they graze, with perhaps 20 to 30 meters between small groups.

By May, the rains are coming to an end, grass growth is slowing, and the herds begin to move westward into long grass. This is the time when you see the iconic, dense herds so familiar from films. The wildebeest bunch up, standing shoulder to shoulder, their movements slowed down by two things: first, the tall grass takes longer to eat, so the front of the herd, often 20 kilometers wide, slows down while those at the back, with less to eat, move faster and catch up. Second, there is danger in that long grass—lions are masterful at concealing themselves and waiting for the wildebeest to walk into them—so the front of the herd moves cautiously. Quite often, a narrow band of zebra walks in front of the wildebeest. You might wonder why zebra put themselves in the most dangerous place. In fact, they are being clever. Our predator biologists, George Schaller, Brian Bertram, Craig Packer and others,[4] have found that wildebeest are the preferred food for lions, but if wildebeest are not around, zebra are next in line. So a sensible zebra makes sure that a wildebeest is indeed around, as close as possible. Zebra stand right next to wildebeest during these movements (and only then). Because of their higher nutritional requirements, zebra need much more grass than wildebeest, so they cannot stand at the back of the herds because there is nothing to eat. Nor can they be in the middle, because there is too much disturbance: May to June is the time that the wildebeest rut takes place. The mating season lasts only a few weeks, but during this time the males set up temporary territories of perhaps 50 meters across and spend their energies in a frenzy of fighting or rounding up females (which take no notice of them, moving in and out of territories as they feel like it). This creates chaos in the middle of the herds, and zebras avoid it because they tend to get rounded up too.[5] Thus, the only place for zebras to go is in front.

Those dense herds on the western plains are seen only briefly during the migration cycle. After a few days, the front reaches the woodlands and the huge herds break up into smaller groups, perhaps 10,000 in each, that disperse into the savanna. This is the early dry season, June–July, and the herds move west past Moru Kopjes and Seronera and along the western Corridor, following the Mbalageti and Grumeti rivers, although some move directly north from the plains past Banagi and Lobo Kopje. The herds move slowly; there is still plenty of green grass, but they also need drinking water, which they get from the rivers, and some are caught by crocodiles in the Grumeti. July is when most of the grass fires occur, set by both poachers and park rangers, and this reduces the food for the animals, which now must move on. Some move outside the park, north of the western Grumeti, where there is more food because cattle grazing prevents burning, although here they become exposed to poaching. Most wildebeest, however, turn north and reach the Mara River by August. Only a few remain in the corridor for the whole dry season. (There is a separate group of about 15,000 wildebeest that stay in the far west, near Lake Victoria, all year—they look different and have a different breeding season.)

The late dry season, August–October, sees the majority of the wildebeest in the far north, above Lobo Kopje and the northern Grumeti,

and centered on the Mara River in both Serengeti and the Mara Reserve, in Kenya. This is the season when food becomes limiting (see
chapter 4), so animals move around a lot, following the small rain show-
ers that still occur in this region. The area is at the high-rainfall end of
the rainfall gradient (chapter 1). As noted, the wildebeest rely predominantly on the rivers for drinking water, particularly the Mara and its
tributaries in Serengeti such as the Bologonja. From time to time there
are mass drownings, several hundred at once, as groups trying to cross
the rivers panic and stampede, some jumping on top of others. More
importantly, this is the season of starvation when many juveniles and
old animals die from disease and lack of food.

Storms become more frequent in October, but the wildebeest do not
move until rain becomes general, usually in November but sometimes
as late as January. When the rains start in earnest, the whole population
moves south at once in groups, often running, and they can cover the
distance to the edge of the plains, some 100 kilometers, in two days.
Here they stop until rain appears on the plains. This is when they are
most spread out, often covering the entire center, west, and south of the
ecosystem. It is also when they are found on the tops of hills—unlikely,
rocky places, but often with young, nutritious grass.

This migration has been taking place for millions of years. The strong
rainfall gradient that determines both the type of vegetation and the migrations of these large herbivores originated some 1.5 million years ago.[6]
In the eons since then long episodes of fluctuating wet and dry periods
have affected where the migrations went, as we will see in chapter 9.

———

This, then, is the seasonal movement of the migration. The wildebeest
are the main species, but they are accompanied by three others: zebra,
eland, and Thomson's gazelle. Zebra move together with wildebeest, usually (for the reasons noted above) ahead of the main groups. Their numbers have remained at about 200,000 for as long as we have counts, even
during the period of increasing wildebeest and buffalo numbers after
rinderpest disappeared, because they were not affected by the disease

(chapter 3). However, about half of the zebra population does not move with the wildebeest but remains spread through the savanna, with perhaps their own movements in the corridor and center of the park, as Grant Hopcraft and Tom Morrison are now discovering. Julian Huxley mentions that zebra had an east–west movement near Lake Victoria before the 1920s.[7]

Eland, of which little has been said so far, also have a migration. A small population of about 20,000 animals moves onto the plains with the other species, being generally mixed feeders—that is, feeding on both grasses and herbs. These are magnificent animals, standing a meter and a half at the shoulder and weighing 500 kilograms or more, and groups of several hundred are a fine sight on the plains, although they are extremely timid and can be seen only at a distance. As the dry season approaches, eland move straight north to the broad-leaved *Terminalia* woodlands around Kogatende to feed on shrubs and young trees; they are entirely browsers in the dry season.

Thomson's gazelle, as we have already seen, remain east of the wildebeest in the wet season, and then follow them west as conditions dry out. They number about 300,000 and benefit from the wildebeests' prior grazing in the long grass. The smaller gazelle (standing 60–70 centimeters at the shoulder and weighing 20–35 kilograms) take advantage of the grazed and downtrodden grass and also make use of the regrowth stimulated by the grazing (chapter 7). However, this lasts only a short while because the gazelle remain in the central woodlands and do not move north. There is a separate population of Thomson's gazelles that live in the Mara Reserve and northern Serengeti.

Grant's gazelle, a larger version of Thomson's, are far less numerous, possibly 30,000–50,000, and are not generally regarded as part of the great migration. Like many gazelle species in Africa and Asia, Grant's are drought adapted; they do not need to drink probably for weeks at a time, and are entirely browsers on the small herbs. They can remain on the waterless plains in the dry season, though we think that many move west to the woodlands' edge. Grant's gazelles are, however, found in groups of 10 to 100 throughout the savanna in both wet and dry conditions.

Perhaps the last species we should mention showing long-distance movements is the elephant. These animals, about 3,000 of them in Serengeti, do not move with the other species; they have their own habits. There are three populations. One is based on the Mara River and spends the dry season in the riverine forests. They move over both the Mara Reserve in Kenya and into Tanzania. In the wet season much of this population moves south into the center of the park along the Grumeti, Orangi, and Nyabogati rivers, where large herds congregate for mating in May. The second population is based in Maswa Game Reserve in the southwest and southern ranges of Serengeti. These animals move east onto the plains and even to the woodlands around Lake Lagarja in the wet season. It is a magnificent sight to see a hundred or so animals marching in single file on the horizon across the plains at this time of year. A third population, which was derived from the northern group in the 1960s, has become resident in the western corridor. At least one group of 300 live around Kirawira, but in recent years they have started to wander, even being found far outside the park, north of the corridor.

In general, the migrating ungulates are following the rain. In the wet season, they are on the far-eastern plains, where, perversely, the rain is least but the grazing is best for animals that like short grass. But these plains dry out first, forcing the animals to move westward. As described in chapter 1, the gradient in rainfall is such that the least falls on the southeast plains and the most in the northwest woodlands. So, as the dry season sets in, the animals simply move up this gradient. Rain is driving the migration in the dry season.[8]

However, this discovery raises the question of why the animals don't stay in the wettest areas all year round, like some of the other nonmigratory species such as buffalo, kongoni, and impala. One possibility is that they are avoiding predators (plains are relatively safe); or they may prefer drier ground. But the most likely explanation seems to be food. The first plant ecologist working on the Serengeti, the Dutchman Hubert Braun, showed in the 1960s that grass on the plains was high in nutrition.[9] Later work by Martyn Murray and others showed that the plains were unusually high in many valuable minerals such as nitrogen, calcium, and phosphorus compared to the dry-season refuges in the north.[10]

FIGURE 6.1. Dung beetles bury the dung of wildebeest and zebra on the plains, and so contribute to the high nitrogen content of grasses that the herbivores favor. December 1999.

Then we have the dung beetles: when wildebeest are on the plains in the wet season, they are followed by dung beetles, hordes of them, laying their eggs in and raising their larvae on the dung. When a wildebeest deposits dung, the flying beetles, about four centimeters long, pick up the scent with their sensitive antennae and home in. It takes only 30 seconds for them to arrive, and they simply dive onto the dung and rapidly fashion it into golf-ball-sized egg containers They roll the balls a meter or so, then bury them with the eggs inside (figure 6.1). The whole dung pile disappears into the ground in a few minutes. The nitrogen in the dung fertilizes the soil, helping it produce the highly nutritious grasses; this high-value food is the reason why wildebeest (and the other species) come to the plains. Dung beetles, therefore, are possible contributing factors to the causes of migration.

Dung beetles can only operate when the soil is soft and wet, hence their action on the plains. In the dry season of northern Serengeti, the soil is rock-hard, and dung beetles cannot burrow and bury dung, so the rapid recycling of nutrients to the soil does not occur and the grasses are correspondingly poor in quality. This is why wildebeest do not remain in the savanna in the wet season. Nutrients do eventually reach the soil through the action of termites, which take the place of dung beetles, but the nutrients are concentrated in the immediate vicinity of termite mounds rather than spread evenly across the landscape, and it takes a lot longer.[11]

With all this knowledge about rain and food, I realized that migration might be an adaptation to make use of temporary good food. If animals can move, they can go to the best food in the ecosystem, fatten up, and then move back again when conditions become too difficult to stay there. Only those species that can eat that type of food would benefit from migration.[12] To test this theory of migration, we needed to study another system.

———

Here we meet the irrepressible John Fryxell, a tall, cavalier extrovert with a great sense of humor and a brilliant mathematical brain.[13] He joined me in 1980 to study migration and has continued to work with me ever since. An opportunity arose for us in the late 1970s, when another great migration of mammals was discovered in southern Sudan. Civil war in Sudan had been raging through the 1960s, but a peace treaty of sorts had been agreed to, in 1972, between the Arab north and the African south. The calm that followed allowed conservationists to survey the country. This migration involved a beautiful antelope, the white-eared kob, which likes to live in wet plains and swamps. The Wildlife Conservation Society of New York was interested in finding out more about this mysterious species—how many were there, where did they go, how threatened by extinction were they? We did know that their numbers were huge and that they traveled a long way. It seemed that this was a migration similar to that in Serengeti, an astounding find if true. It was also an opportunity to see if our understanding of how migrations worked, derived from the Serengeti research, held true in this one. John Fryxell took on the task.

Sudan is remote, out of touch with the modern world, and cut off from supplies. The southeast corner of Sudan, called the Boma, is a vast area of flat grassland, scorching hot, an impassable quagmire in the rains, and bone-jarringly rough in the dry season. The soil is of silt, locally called "black cotton," which bakes hard as rock in the dry season and breaks up, producing cracks wide enough to take half a wheel. Progress is a painful 10 kilometers per hour, destroying vehicles and causing them to overheat. At the first heavy rain this silt turns immediately to a bottomless glue, and vehicle progress slows to zero. Walking also becomes impossible, as the gumbo accumulates on boots with every step so that after 10 steps the weight is too heavy to raise one's foot. It is little wonder that wildlife still survives—no one can get in to destroy it. The puzzle is how the wildlife manage to survive in such physically demanding circumstances—and there is much wildlife, for besides the white-eared kob, there are also good numbers of hartebeest, topi, zebra, Thomson's gazelle, giraffe, elephant, and even a few herds of buffalo (at least, there were in the 1980s). In short, the "gumbo" protects the wildlife despite the human conflicts.

John Fryxell spent three years in this remote place, following the kob to the open grasslands in the south during the wet season. The grass was still long by Serengeti standards, 50–100 centimeters high, but it was green and the ground was firm. As conditions dried, the kob moved 70 kilometers north to the swamps, which remained wet even at the worst times—all 900,000 of them lived around the swamp edge. This was the refuge. John measured the amount of food and its nutritional value. His findings showed a remarkable similarity to the picture we had developed for Serengeti.

John then used his research in Boma to develop the general ideas about migration, the reasons they occur and how they work.[14] In essence, migration occurs when animals move to take advantage of temporary sources of good food, but then have to move away again when those sources disappear. The extra food that they obtain from this behavior allows large populations. We came to realize that all migratory populations of large mammals—caribou in Canada, bison in North America, numerous examples in Africa, and even whales in the oceans— can support large numbers because they have access to this additional high-quality but temporary food.[15] Migratory birds are probably also taking advantage of these kinds of temporary food supplies when they migrate to the high Arctic for the northern summer.

———

Are the movements of our long-distance migrants really different from those of resident animals? Migrations are driven by the seasonal environment. Are resident species similarly influenced? Let us first look at the resident herbivores in Serengeti. Impala live in acacia savanna. Their dry-season refuge is in thickets and denser vegetation close to rivers; in the wet season they move upslope perhaps half a kilometer onto the tops of ridges, where it is more open. Buffalo, topi, and kongoni show the same movements away from rivers in the rainy season. Leopard tortoises live in the dense, riverine grassland next to rivers in the dry season but move out a few hundred meters onto rocky hills when the riverine areas become swamped. In other parts of the world, we see moose in Sweden living close to the coastal valleys in winter (their refuge) but moving up to mountaintops in summer; elk do the same thing in Banff National Park, Canada. Birds show similar altitudinal movements: in Australia, honeyeaters near the coast in winter move to the mountains in summer, as do some birds in Tanzania, moving from the coast in the cool season to the Southern Highlands in the warm season. These few examples reveal that the element common to so-called residents and long-distance migrants alike is that they all have refuges and different locations where food is abundant, and they all move between them, driven by the seasonality of the environment. The conclusion we draw is that where there are seasons then there are movements—and there is no place on Earth that is not seasonal. Even in the deepest parts of tropical forests, or in deep oceans, there is seasonality. From this aspect, the only difference between residents and migrants is one of scale; residents do not move as far as long-distance migrants.

Considering this, is there anything special about long-distance migrants? Yes, there is. They have two adaptations that make them different.[16] The first is their ability to find and eat the temporary but abundant good-season food. Buffalo cannot migrate onto the Serengeti plains because they cannot eat the short grass there, but wildebeest and others can. Insectivorous birds in Serengeti migrate to Asia or even to the Arctic to eat the superabundant insects in summer, but seed-eating birds cannot do that because the same seasonal supply of seeds does not exist. Plankton-eating whales migrate to the Arctic to eat the abundant summer ocean krill before returning south in winter (often not eating at all but living on their fat supplies), but carnivorous whales cannot migrate that far because there is simply not the same abundance of their food. The 8 million straw-colored fruit bats living in the dense tropical forests of west and central Africa migrate more than 2,000 kilometers to a small reserve in Zambia (Kasanka National Park) to take advantage of an abundant, rich, but temporary supply of fruits from three species of forest tree. These trees fruit for a few weeks at the beginning of the rains in November. Once this temporary glut is over, the bats disperse and make the 2,000-kilometer return trip across the Congo. Insectivorous bats do not show the same movement.

The second adaptation peculiar to migrating animals is their ability to move long distances. Wildebeest have special pogo-stick-like springs in their ankles that allow low-energy, long-distance running;[17] impala have wonderful abilities for jumping but not for long-distance walking. Early humans evolved bipedal walking, an energetically efficient mode of long-distance travel, which would have allowed them to follow the migrations for the purpose of scavenging (a huge source of temporary food—10 times that from resident prey); humans are the only scavenger/predator capable of doing this. Carnivores cannot migrate with their prey because their babies cannot walk far for at least two months, whereas wildebeest babies can run as fast as their mothers within 24 hours. So long-distance migration is only possible for those few species with the appropriate adaptations.

————

Given that we have a migration of a vast population of animals, what are the consequences of this phenomenon for the ecosystem? First, as mentioned earlier, there is a difference along the rainfall gradient in the fertilizing of the soils by the dung from the migrants, and this spatial difference affects the grasses and the animals that live in those plant communities. Second, and perhaps most importantly, the migration creates stability in the plant community and prevents overgrazing in the system. Overgrazing occurs when large herbivores such as cattle graze on grasses so heavily and persistently that the grasses can never grow up, flower, set seed, and senesce. (Senescence is the process that grasses perform in which they take the energy and nutrients they have built and send them to their roots for storage over the dry season, to be ready for the next rains.) If grasses cannot store their food, they die after a year or two. To ensure adequate storage, all grasses require a period of rest from grazing to allow them to build their food stores, just as we saw wildebeest depositing their fat stores (chapter 4). Migration allows just this rest period in all areas of the park. On the plains, wildebeest depart before the grasses dry out, leaving time for flowering and storage, which in turn increases soil nitrogen and future growth.[18] On the savanna, wildebeest are not usually present during the wet season, so grazing does not prevent flowering. In the dry season, the migrants eat what is above ground—the dry material plus a bit of regrowth—but they cannot damage the grasses' food stores in the roots. This means that overgrazing does not occur even though wildebeest eat effectively all their aboveground food; the system is stable both on the plains and on the savanna.

The migration also determines the plant-community complex on the plains: a community of short grasses and low, creeping herbs that disappear when the migrants disappear, as they did in the 1930s. The migration also affects insects, as we saw with the dung beetles, and indirectly affects a number of bird species such as the huge kori bustards and the black-winged plovers, gull-billed terns, and Eurasian rollers that follow the wildebeest, because they all eat the beetles. The communities of insects and birds that live on the short-grass plains are dependent on the migration to maintain their habitat.

Predators are also significantly influenced by the migration. Wildebeest calving takes place in February, when wildebeest are normally on the short-grass plains. The newborn calves are extremely vulnerable for a few days, are easy to catch, and all predators move onto the plains to eat them. Lion prides that live at the edge of the savanna move their territories some 50 kilometers. Hyena packs that live on the edge also move farther onto the plains. Cheetah follow the gazelle out to the plains (they also catch baby wildebeest), and wild dogs (when they were present in the 1960s) also made their dens on the plains in this season. Both golden wolves and silver-backed jackals benefit from the new food. This is the time of plenty for these predators. Conversely, the same species in northern Serengeti have little to eat; this is their lean time because they cannot migrate along with the wildebeest—they all have territories and must stay in them. Hyena clans set up their dens on the edge of the plains and commute up to 50 kilometers into the corridor to reach the wildebeest, but beyond that is too far for commuting, as Hans Kruuk found in his detailed studies in the 1960s.[19] Hyenas in the savanna are often found as single animals rather than in groups. But whether they hunt singly or in groups, none of the large predators can follow the migrant herbivores' seasonal movements the whole way through the year. The reason is that newborn lions, hyenas, and other predators cannot move far (figure 6.2). They must spend some weeks growing before they can travel, forcing the adults to remain in one place. For this reason, their populations are determined by what other resident prey are available.

The top predator populations (lion and hyena) are bottom-up regulated by resident prey. However, there is a twist to this story. As the wildebeest population increased in the 1960s, so did the number of old and decrepit wildebeest that could not keep up with the migration; they stayed behind, alone and vulnerable. As wildebeest numbers rose, they became an increasingly significant part of the food supply of lions, as Grant Hopcraft and Craig Packer found, so that lion numbers increased with wildebeest numbers.[20] In effect, the bottom-up regulation of predator numbers was inflated by the migration, and these extra predators had a corresponding depressing effect on the top-down-regulated resident herbivores (see chapter 5). The migration, by its large numbers, modified both the bottom-up and top-down interactions between predators and prey.

FIGURE 6.2. Lions cannot migrate with the wildebeest because they have to carry their
newborn young. Mara Reserve, May 1982.

Consequently, we can say this about migration: long-distance movements result in bottom-up regulation of the wildebeest, with only
25 percent of deaths due to predators, compared to top-down regulation
of resident animals, with some 80 percent of mortality from predators.
Migration, therefore, allows considerably higher populations due to
both extra food and escape from predation, neither of which are available to sedentary populations. In turn, these numbers change the bottom-up regulation of plants to top-down regulation by herbivores. This
answers my starting question: What is so special about migration? In
two words, higher populations.

Therefore, the consequence of migration is that it becomes a dominant structuring force in ecosystems, what is called a *keystone effect* (see
chapter 7). This is only too clearly seen with the wildebeest, whose huge
numbers, resulting from migration, affect every other aspect of the
Serengeti ecosystem.

7

Biodiversity and the Regulation of Ecosystems

Processes That Create Biodiversity

In 1992, the nations of the world met in Rio de Janeiro and set up the Convention on Biological Diversity.[1] This was the first time that the conservation of biodiversity was recognized as an important, even integral, part of human development. Tanzania was a signatory and as a result was committed to documenting and protecting its biota. In Serengeti, I realized, this begged a number of important questions. In a square meter of soil there are thousands of bacteria, fungi, soil animals, plants, and insects. Now broaden the scale to 10 square kilometers and add in more plants, insects, birds, and mammals. We knew what regulated the numbers of a few mammal species, but what determined the numbers of all these others? What made the Serengeti so biodiverse? Why were there so many species? More importantly, were all these species' populations independent of each other, or were they somehow connected as part of something like a superorganism?

———

We begin by asking what are the processes that create biodiversity in Serengeti, and allow so many species to coexist? One possible organizing mechanism is the creation of different niches for different species through competition among the species. Given that there is not enough food to go around, the different species must compete with each other. Because the species are different, they have unequal food-gathering abilities—for example, small species can eat small food items more efficiently than can large species, and vice versa. Over evolutionary time these differences are selected for, and the species become even more different. The type of food, habitat, and environment in which each survives best is called a *niche*. By living in separate niches, the species still compete with each other, but they are kept at a level where all survive. The important point is that resources (food or space) must be in short supply at some point in the year, compelling species to compete. Some have chosen habitats that are extensive, like the wildebeest on the short grasslands, so that they can be numerous. Others, like the klipspringer, a small antelope, live on top of kopjes, which are few, so their numbers are very low. We might say that the relative abundance of an herbivore species (wildebeest common, klipspringer always rare) is in all cases dictated by its resources.

The early research in Africa, particularly that of Hugh Lamprey in the late 1950s, showed that different ungulates lived in different habitats. Some liked dense thickets (kudu, dikdik), others more open woodland (impala), others swamps (lechwe, sitatunga), while yet others preferred long grass near rivers (buffalo, waterbuck, reedbuck) or short grass on ridges (wildebeest). They all differed, and they all had different-shaped mouths capable of eating different food—tips at the ends of tree branches (giraffe), coarse woody shrubs and trees (elephant), long grass (buffalo, topi), short grass (gazelles).[2] Richard Bell, one of the early scientists in Serengeti, and Martyn Murray later in the 1970s and '80s, showed that grass eaters actually chose different bits of the grass plant—zebra liked coarse stems, topi preferred leaves high up on stems, wildebeest the leaves on the ground. More recently, DNA has been used to document the types of food that species eat.[3] The use of camera traps—cameras set up on posts that are triggered to take photos when animals move past—confirms the earlier finding that different herbivores prefer different habitats. However, there is still considerable overlap in the habitats they use.[4] Some species also differed in the environments they liked: kongoni and topi are closely related antelopes, similar in size, and both prefer to eat similar tall grass, but they differ in their environments. Topi like the wet climates of the west, kongoni the dry climates of the east, with a narrow area of overlap in the middle. Of the tiny antelopes, dikdik prefer dry conditions, steinbuck intermediate climates, and grey duiker very wet conditions. By 1979, I had concluded, in our first synthesis of the Serengeti research,[5] that competition was driving biodiversity in this ecosystem. It turned out that this was only part of the story.

———

In the mid-1960s Robert Paine, of the University of Washington, came up with the ground-breaking idea that not all species in a community are equally important in keeping the community together; some are more important than others, and he called these *keystone species*. Crucially, these keystones are at the top of the food chain—the predators— and they have the effect of stopping any one prey species from becoming dominant and excluding others. Paine studied this process with starfish feeding on mussels, barnacles, and other species along the intertidal zone of the Pacific Coast, devising an experiment in which he removed starfish for a year or so and watched what happened. He found that mussels took over the rocky shore and excluded many other smaller species.[6] The classic example of a keystone species, the Pacific sea otter, was demonstrated by Jim Estes, working in Alaska: when 99 percent of the otter population was removed by the fur trade in the 1800s, sea urchins, the prey of the otters, rapidly increased. This resulted in the loss of giant-kelp stands and their inshore fish communities, replaced by sea-urchin-dominated grazing lawns of small algae.[7]

Keystones created biodiversity—a fundamental insight. While I worked with Bob Paine in the 1970s, I realized that keystones need not always be top predators. The Serengeti wildebeest could act as a keystone because it determined the diversity of most other parts of the food chain through top-down processes (chapter 5, also figure 7.1). And elephants could also act as keystones on other parts of the community of species, particularly the plants and other herbivores.

FIGURE 7.1. Wildebeest herds affect everything in the Serengeti. They are the keystone species. April 1973.

It was not just competition that determined the coexistence of species, as I once thought. I now saw that two processes likely create biodiversity: competition and predation.[8] Competition for food and predation not only regulate populations of species (as discussed in chapters 4 and 5), but also allow many species to coexist in a community. But this is still only part of the answer, because there is a yet more complex outcome of the combined influences of competition and predation, and this involves the evolution of social behavior. The remarkable work of Peter Jarman was the first to document the adaptations of the African ungulates in a classic treatise in 1974.[9] Jarman pointed out that the size of a species dictates both the type of food it can eat and, simultaneously, the ways it can avoid predators.

A small species of herbivore does not have to eat a lot, but what it eats must be high in protein and other nutrients. It turns out to be a constraint of nature that high-quality foods are scarce in proportion to the time that an animal must spend searching for them. These animals must be solitary because there is simply not enough food to support a bigger group. In contrast, large species have the opposite requirement: they need a lot to eat, but it does not have to be high in quality. In nature this describes the majority of the vegetation—lots of low-quality, tall, rank, coarse grass and woody shrubs and trees. And because this food is plentiful these species can live in groups or herds. For example, the tiny dikdik feeds on buds and stem shoots of small bushes, which it must search for and pick off; as a consequence, it must live solitarily or in pairs. The small Thomson's gazelle, which eats grass, needs the high-quality new shoots at the base of growing plants; these are never in great supply but are more plentiful than the dikdik's food, so the gazelles can live in scattered groups. At the other end of the size spectrum, wildebeest, zebra, buffalo, topi, and other larger antelope, which feed on abundant grass, are able to live in dense herds.

All of these species have evolved different behaviors to avoid predators. Small, solitary species avoid predators by hiding. Duiker, steinbuck, oribi, and dikdik rely on camouflage and secretive behavior. Larger species that live in groups would find no benefit in trying to hide, as they are too easily detected by predators; instead, they rely on safety in numbers. When a predator attacks, the many eyes and ears can detect it more efficiently, and a predator can (usually) only catch one of the group. This lowers the chance that any individual will be caught (though bad luck for the one that becomes dinner).

A species' social behavior in turn dictates its mating behavior. Solitary species set up permanent territories, which males defend to preserve food supplies and attract females. Large species like zebra, buffalo, and elephant cannot set up individual territories on a permanent basis since they live in groups that move around. In these species, a hierarchy among males develops in which males fight and the top winners then have rights to every female. Some medium-sized ungulates, like wildebeest, topi, and impala, set up temporary territories lasting for as short a time as a day and hope that receptive females pass through at the right time.

These complex adaptations to the combination of competition and predation, so elegantly described by Jarman, contribute to the diversity of herbivorous mammal species in the community. They provide one answer to our original question: How can there be so many species?

Does Biodiversity Strengthen Stability?

Now that we have seen how biodiversity is created, this leads to the next question: Does biodiversity strengthen regulation? That is, does it result in greater constancy (meaning less variation) of population size or a greater constancy in the number of species—what we call *stability*—and a buffering of disturbance from the environment (disturbance is discussed in chapter 8)—what we call *resilience*?

There has been considerable debate as to whether or not diversity helps to maintain and stabilize natural ecosystems.[10] If the presence of many species does help to stabilize an ecosystem, then the loss of species might result in its rapid change to another combination of species, a different state (for discussion of multiple states, see chapter 10). This issue has had obvious significance not only for conservation but also for the persistence of human systems, because when humans convert natural habitats to agriculture or other land uses, they reduce the natural diversity significantly, something we will return to in chapter 13.

To answer these questions, we need two bits of evidence. First, we ask: Does the presence of some species allow, or even enhance, the survival of other species? If so, this means that species depend on each other, and translates as increased constancy of the diversity of species and of species population sizes. But there is another way that biodiversity can lead to stability: with more species, a drop in population size of one species can be balanced by an increase in the population of another. This phenomenon is called *compensation*—one species compensates for another, so although there is high variability in numbers between species, the overall total of individuals is more constant with more species in the community (because there is a greater chance that one can replace another). If the answer to any of these aspects is yes, then we ask the second question: What are the ways that species depend on each other to produce such stability of number of species or total number of individuals in the community?

To answer the first question, we need to see either that the removal of one or several species leads to a decline in numbers or even complete loss of one or more other species (and sometimes may indirectly cause an increase in others), or, conversely, the addition of one species leads to the increase of many others too, although yet others may decrease because of indirect connections. Second, we need evidence for compensation.

In the Serengeti we can use the wildebeest as a dominant species whose numbers have changed. We do not have any information on what happened within Serengeti when rinderpest arrived in 1890 and reduced wildebeest numbers to low levels, but we do know what happened when they came back: we saw that many other species were affected. Some top predators increased, which resulted in other predators decreasing. In turn, several resident herbivores were regulated by predation due to the increase in these predators (the evidence from the removal and subsequent return of predators in northern Serengeti is proof of that; see figures 5.2 and 5.3). The increase in grazing changed the grasslands on the plains from long to short, and with this came a change in bird communities (Ally Nkwabi's studies of fire and grazing demonstrated that[11]). The increase in grazing and decline in fire increased both tree numbers and diversity, as well as all the species associated with trees.[12] Thus we have evidence that an increase in a keystone species results in an increase in number of other species.

Was there greater constancy of abundance of each prey species when predator numbers increased? We can see from the relative changes in numbers of both impala and topi in northern Serengeti that they were far more stable (less variable in numbers) with predators present (1967–1972) than when predators were absent (1980–1993).[13] On the larger scale, once the wildebeest population had leveled out, we saw that the relative population sizes of the grazing-mammal species remained the same for the next 40 years. There are always about 1.3 million wildebeest (once they had recovered from disturbance) and 8,000 kongoni, the latter roughly 150 times fewer in number but very similar in ecology.

However, instead of greater constancy of individual species numbers, is there evidence of greater constancy of total community abundance as a result of compensation (despite high variability of individual species populations)? We have evidence from bird communities: when grasslands change from long to short grass, the bird species change but total numbers remain similar because as one species declines in abundance, another increases.[14]

So sometimes we see a greater constancy of abundance of each prey species when biodiversity increases, and sometimes we see high variability in numbers between species but a constancy of overall abundance as biodiversity increases (due to compensation).[15]

What Are the Ways That Species Depend on Each Other?

How is stability achieved? If more species lead to more stability in the community—either via constancy of species numbers or constancy of total population abundance—this suggests that species must be depending on each other. What are the ways that species are linked together in the Serengeti?

By the 1990s I had found in studying the Serengeti buffalo and wildebeest that individual species' populations are stabilized through changes in births and deaths (what we call *feedback mechanisms*; see chapter 3), work that was later corroborated by Herbert Prins in other protected areas. But a community is composed of thousands of species, and if they all operated independently of each other the community should change rapidly as some species drop out and others come in. Likewise, the relative proportions of species should change markedly over time. Yet this does not appear to happen. Research at Olduvai Gorge by Charles Peters and Robert Blumenschine of Rutgers University has shown that over the past 4 million years the number of carnivore species in the Serengeti ecosystem has remained between 9 and 19, and in the last 2 million it has remained between 9 and 14 species.[16] The types of species have evolved, with new forms replacing old forms of similar ecology so that the number remains relatively constant. The question then becomes, Is there a higher-order principle that explains not just how populations of individual species are regulated, but how whole communities of species are regulated? Is there a special connection among the different species that creates and shapes the community we see in Serengeti—or any other part of the world?

———

By the late 1990s, information was coming in not only on the regulation of large and small herbivores, but also on the diets of carnivores, through the work of our colleagues Craig Packer, Meggan Craft, and many others on lions, Hans Kruuk and Heribert Hofer on hyenas, Brian Bertram on leopards, and various other scientists looking at the smaller carnivores. Armed with this expanded knowledge, I was struck by the relationship between the pattern of death in herbivores and the diversity of carnivores.

As we saw earlier, top-down or bottom-up regulation of resident ungulates is determined by body size (chapters 4 and 5). Very large ungulates such as giraffe, buffalo, and other larger species have outgrown their predators so that predation is a minor cause of death in adults. In general, mammal herbivores over 400 kilograms (the size of buffalo or larger) in Serengeti suffer little or no predation as adults and are regulated through food supply. This was confirmed in other areas, such as Kruger National Park in South Africa, where the larger ungulates were also food limited.[17]

Below a certain body size (about 150 kilograms, the size of a kongoni or wildebeest), there is a rapid switch from bottom-up to top-down regulation. This occurs because as prey become smaller they are exposed to a greater number of predator species until below a particular size, all adult deaths are caused by predation, as occurs in species such as impala (50 kilograms) and oribi (10 kilograms). Figure 7.2 shows how the increased number of predators of small prey results from the food niches of small predators lying within those of larger predators.[18] This illustrates the fact that large predators like lions can eat small prey like oribi and even hares, but small predators cannot eat large prey. Although the average size of preferred prey is different for each predator, the overlap

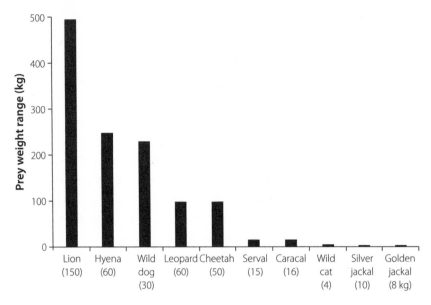

FIGURE 7.2. The breadth of diet of a predator is related to its body size. A lion can eat a larger array of prey sizes than a wild cat can. (Data from Sinclair, et al. 2003.)

in niche results in much higher incidental predation of smaller prey compared to that of larger prey. What this means is that *both* the diversity of predator species and the diversity of prey species produce this pattern of regulation. If the diversity of either prey or predators is reduced, the pattern would disappear. In essence, the size diversity of both predators and prey results in both top-down and bottom-up processes occurring in the same system, and this dual process results in increased stability of the system. Thus, diversity may lead to stability, as we now explain.

The likely stability of this combined suite of prey and predator species results from the way food niches of smaller species are embedded within those of larger species. Overlap among the dietary niches of predators, shown in figure 7.2, determines what happens when a top predator is removed from the ecosystem (for example, through poaching or other disturbances, discussed in chapter 12). This top predator is often replaced by a smaller predator species—recall in chapter 5 how leopards increased when lions were removed in Ngorongoro—a change that is called in the jargon *mesopredator release*. Mesopredators, which means "smaller predators," are limited in numbers because top predators kill them, or they are forced into marginal habitats to avoid the top predators. So in the absence of top predators, the increase in these smaller predators then hits the smaller prey and reduces their numbers. Justin Brasheres, one of our team in the 1990s, found in Ghana that removal of leopards allowed a 300 percent increase in baboons, which in turn reduced numbers of many small ungulates, including oribi and small monkeys.[19] Similar changes in North America have followed removal and reintroduction of larger predators such as wolves.[20] Thus, somewhat counterintuitively, the presence of top predators makes it possible for smaller herbivore prey to persist in higher numbers. The important point concerning biodiversity is that we have a range of predator species and a range of prey species that together result in a stable community resistant to disturbances—they create resilience in the system. Reduction of these size ranges is likely to result in instability of the system.[21]

—————

Species can be linked by a special kind of predation. These involve the specialist predators—those that are entirely dependent on one or a few prey species, so that loss of the prey species results in the loss of the predator. They differ from most predators, which normally have a wide range of prey types so that loss of one has little impact on the predator. Among Serengeti mammals the aardwolf feeds largely on one genus of termite, *Trinervitermes*, which account for only about 30 percent of termite mounds but are vital to the aardwolf. Loss of this one genus of termite would mean a severe decline of aardwolf.[22] In the insect world, effectively every species of butterfly or moth has a corresponding parasitic wasp (called a *parasitoid*), which lays its eggs on the caterpillar; its larvae burrow into the caterpillar, usually allowing the host to pupate before eating the host inside out so that at hatching the adult wasp appears, not the moth. Each parasite is specific to the host. The extreme significance of this is that parasitoids, remarkably, make up 25 percent of all insects.

The top-down effects of keystone species such as wildebeest and elephants, mentioned above, result in many other species becoming dependent on the keystones. We saw how this happens when the keystone (wildebeest) numbers change. Elephants are another example. The role of elephant as a keystone was first described by the remarkable naturalist Desmond Vesey-Fitzgerald in 1960. He recognized that elephants create niches for other large grazers in the swamp grasslands of Lake Rukwa, in southern Tanzania. Swamp grasses grow tall, up to three meters. Elephants have no trouble eating and trampling these, thus creating open patches where the flattened grass can sprout succulent shoots half a meter tall. These areas opened by elephants are the perfect niche for buffalo, which graze the grass down to a shorter height. In turn these grazed swards are used by waterbuck, a large antelope that likes wet areas. Vesey called this sequence, in which one species creates a niche for another, a *grazing succession*.[23] It is a process that is recognized in ecology as *facilitation*: one species provides a feeding niche for a second species.

Richard Bell in the 1960s examined whether Vesey's theory applied to Serengeti. He recorded a grazing succession among resident herbivores in western Serengeti.[24] Buffalo, zebra, wildebeest, topi, and Thomson's gazelle shared that area. In the wet season they all grazed next to each other on the tops of shallow ridges where the grass was short and nutritious. Downslope the grasses grew taller and more fibrous, reaching over a meter in height in valley bottoms near rivers. As the grassland on the ridges dried out at the end of the rains, the larger animals moved downslope first. The sequence was buffalo, followed by zebra, then wildebeest and topi traveling together, and finally Thomson's gazelle. As the species moved down, each grazed the grasses lower, providing a niche for the following species. Sure enough, Vesey's discovery applied to Serengeti.

This discovery raised the question of whether there is grazing facilitation among the great migrating herds of zebra, wildebeest, and gazelle on the main Serengeti plains. Bell speculated that facilitation may be one of the factors allowing the migration and hence one of the reasons that Serengeti is unique. There was some intriguing evidence in its favor. Sam McNaughton, our grassland ecologist, and his team of Joy Belsky and Mike Coughenour had discovered in 1976 that when wildebeest moved off the plains at the end of the rains, their grazing stimulated the regrowth of certain grass and sedge species, producing an additional amount of short, nutritious leaves that Thomson's gazelle preferred. Because the gazelles followed where wildebeest had grazed, it was assumed that they benefited from this growth—a form of grazing facilitation.[25]

The facilitation theory for the migration could be tested. Since we knew the wildebeest were increasing, we could predict that the gazelle population should also be increasing because of the extra food that wildebeest were creating for them. The opposite prediction prevailed if interspecific competition was occurring; gazelle numbers should be decreasing. Through the 1990s both wildebeest and Thomson's gazelle populations were counted, and the gazelle numbers declined to half what they were in the 1970s. Trends suggested that numbers had declined once wildebeest reached their stable level in 1977—the opposite of what we had predicted with the facilitation theory. The main reason was that although gazelle may have benefited from wildebeest grazing in the month or so at the end of the rains (May–June), they subsequently suffered higher grazing competition with wildebeest in the dry season, counteracting the earlier benefits.[26]

———

Facilitation is a type of species dependence, which we call *commensalism*, in which one species benefits another by giving it protection or by creating other favorable conditions for it to live without getting benefit itself. Among plants, the tall red oats grass, *Themeda triandra*, which is normally heavily grazed, obtains protection when it grows near patches of the turpentine grass (sometimes called lemon grass) *Cymbopogon excavatus* on the long-grass plains. Presumably the grazers find it unprofitable to waste time sorting one from the other and move on to other areas. Similarly, in northern Kenya plants use each other to hide from herbivores.[27] The candelabra euphorbia tree (*Euphorbium ingens*), as noted in chapter 1, is a succulent that grows to a large tree, six to eight meters tall. Recall that as a small plant it is easily killed by fire and cannot grow up on its own in the savanna. Its adaptation is to start life at the base of a small acacia tree, usually the umbrella tree (*Vachellia tortilis*), which is bushy and thorny and protects the baby candelabra from fire. This baby then grows thin and straight up through the branches of the acacia until it emerges at the top and spreads out into its classic mature candelabra shape. However, there is a twist to this story: it appears that the acacia itself then dies before it reaches maturity, suggesting that the candelabra kills it once it is no longer needed—the candelabra becomes a competitor.[28] But the acacia does not die in vain; it has increased the nutrient content of grasses growing under it, which benefits savanna grazers.[29]

Among animals, cattle egrets take advantage of the insects disturbed by grazers, as do wattled starlings. Dung beetles follow wildebeest on the plains, but the huge kori bustard and other bird species also follow wildebeest because, as we saw in chapter 5, these birds eat the dung beetles. Some birds, such as the superb starling, nest in the base of raptor nests for protection, and many birds, such as barbets and sparrows, nest in holes created by woodpeckers. One of our students, Ephraim Mwangamo, showed that Fischer's lovebirds (a type of small parrot) find nightly refuge in the roosting nests of rufous-tailed weavers. The parrots wait until dark, when the weavers are in their nests, then dart in unobserved as uninvited lodgers.[30] Birds thus depend on other birds, examples of commensalism.

———

In the above examples of commensalism, one species benefits and the other derives neither benefit nor damage. But there are many examples where both species benefit. This is called *mutualism*. Mycorryhzal fungi, which can live only in the roots of grasses on the plains and provide nutrients to them, allow those grasses to better withstand grazing.[31] With large mammals we see an interesting interaction between olive baboons and impala. The two tend to associate together when out feeding. It appears that baboons, which often have a lookout sitting in a tree to keep guard, provide protection for the impala as well. In turn baboons may obtain benefit from the more acute hearing and smell of impala—although occasionally a large male will kill a newborn impala.

Meredith Palmer has documented the mutualistic association of the two species of oxpeckers with their mammal hosts.[32] Red-billed oxpeckers have scissor-shaped bills and associate with some 18 species of herbivores, on whose bodies the birds find a wide variety of ticks. They roost at night in trees, unlike the yellow-billed oxpecker. This bird has a wide, flat bill and feeds on the large tick species found only on large herbivores, particularly buffalo, giraffe, and eland. Remarkably, the yellow-billed oxpecker roosts on its host, especially the giraffe, under whose belly it will roost for protection. The mammals obviously benefit from the removal of annoying bloodsuckers, but also benefit because the birds' loud, raucous call alerts the host to danger. And the birds, of course, get food. Interestingly, they also attack open wounds on the host, enlarging them and preventing healing. Although this is a minor activity at this moment in evolution, oxpeckers may be well on their way to becoming parasites—perhaps a million years from now they will have evolved into the vampires of the bird world.

Many bird species feed together in groups. We see this with the small insect eaters that live in the tops of acacia trees. It appears that insect food does not occur evenly but in patches, and group hunting can cover a larger area to find those patches of insects. When one bird finds a good patch, it gives a call recognized by the other species. Similarly, different species of small seed-eating birds feed together on the ground. In this case they have a common alarm call so that if one senses danger and calls, the rest respond and escape.[33]

There are many such examples of species that depend on each other. The thin, spiky tree called whistling thorn produces hollow galls that serve as houses for special ant species. The ants defend the tree from browsers by swarming up branches, biting and giving off a pungent aroma—even elephants will abandon browsing after a few minutes, as

Jake Goheen and Tod Palmer have shown.[34] Look around your feet at flowering herbs in the savanna and you will see small blue butterflies pollinating them. There are many species of blue butterflies, most quite common. They all require ants to look after their caterpillars; the ants carry the butterfly larvae into their nests. If we lose the ants, we lose the butterflies, and then we lose the pollination of those plants. Flower pollination by animals is a classic example of evolution for mutual benefit. The characteristics depend on the pollinator. Bird-pollinated flowers lack scent because birds cannot smell, but they can see the flowers' bright-red colors. Likewise, the flowers' long corollas are adapted for long beaks that receive and deposit pollen—and acquire considerable amounts of nectar as a reward. The flowers of *Leonotis nepetifolia* on kopjes and disturbed ground are adapted for pollination by the beautiful sunbird (yes, that is its name). Flowers of baobab trees, which are pollinated by fruit bats, are white and have a strong scent, because bats, like most other mammals, do not see color but have a strong sense of smell. Insect-pollinated flowers have short corollas, less nectar, but strong scent, and are sometimes blue, violet, or even ultraviolet because insects can see ultraviolet light as well as detect scent. Many of the African acacias, such as gum-arabic acacia, are pollinated by butterflies that feed on the nectar.

Embedded food niches, specialist predators, keystone species, commensalism, facilitation, mutualism—these are some of the ways in which species depend on each other as part of the Serengeti community. Beyond the need for other species in lower trophic levels as food, many, if not most, consumer species cannot exist on their own even within their own trophic level; loss of one species can mean loss of one or perhaps many more species.

———

Yet it is not just the number of species but also their social behavior that is important in maintaining stability. Both herbivores and predators exhibit social behaviors that cause some species to live alone and others to live in groups, as Peter Jarman has demonstrated. John Fryxell, whose work on migration we saw in the last chapter, modeled the interaction of predator and prey and found that when predators and prey act as individuals—as single animals—we can get unstable consequences with huge fluctuations in numbers (something we see in northern ecosystems of the world; see chapter 5). But prey that live in herds are harder for predators to catch (because there are more eyes to spot the predator), and consequently the predators are less efficient at capturing prey and so do not cause rapid fluctuations in prey numbers. If the predators also live in groups, as lions do, then the stability is greater still.[35] In general, all these interactions among species—competition, predation, commensalism, mutualism—contribute to how the ecosystem works. Many are special to Serengeti, but many are also general in the world. Kayla Hale and colleagues, for example, have shown that mutualism between plants and pollinators, in particular, strongly enhances stability and persistence of communities.[36] These studies, therefore, provided the answer to my question at the start of this chapter: Is there a higher-order organization among many species that results in greater stability of the ecosystem? It appears that this is indeed the case.

———

For Serengeti, however, the presence of many species together likely explains the patterns and processes in the ecology of the system. The role of biodiversity shows a far more profound complexity than appears from a simple counting of species. We posed the question of whether we need all these species, or to put it another way, do these species depend on each other, require each other, for the system to operate and persist? The answer is unequivocally yes. Species cannot live on their own; they must have others to support them. Some species are essential, the keystones; others are not essential, but if we lose too many of these the system collapses, so together all are important. To conserve healthy ecosystems, we need to protect all their biodiversity. This is an important conclusion for conservation.

8

Disturbance and the Persistence of Ecosystems

I saw the Serengeti plains from the rim of the Ngorongoro caldera in December 1961. They were under water, glistening in the evening light. The migration had disappeared to the highest ground the animals could find, near the Gol Mountains in the northeast. The national park was closed, roads were under water and turned to mud. The Ngorongoro crater itself was so full that the roads were mere causeways in a vast lake. The wildebeest and zebra perched on the sides of the crater to get out of the water. Lake Victoria came up so high that it cut off one of its peninsulas and created the island of Ukerewe, which has remained an island ever since. I did not see Serengeti like this again until 2019–2020, when we experienced the greatest floods since records have been kept. Floods also occurred from November 1997 to February 1998 with similar but not so extensive consequences. In contrast, the rains failed entirely in 1993 and produced the most severe drought of the twentieth century. Since 1961 drought events have occurred several times in different areas of the ecosystem, but only occasionally in all areas in the same year.[1]

———

Research on how the Serengeti ecosystem works has so far focused on regulation and its two main drivers, starvation and predation. The key consequence of regulation is that it stops populations from increasing;

they reach an equilibrium, a steady state, where the birth rate equals the death rate. But this idea of an equilibrium creates an enormous problem: we almost never see such a steady state in nature. Numbers bounce around from year to year and almost never settle down to a steady state—the zebra and wildebeest populations are rare examples of visible stability. In fact the wildebeest population of Serengeti, which has settled down to a relatively steady state, is often cited not so much to demonstrate the equilibrium but because it is such a rare event. In most species, numbers fluctuate a lot, going up or down due to changes in the weather; one never sees the equilibrium, but it is indeed there. The important question then is, how do these fluctuations in environment and populations relate to the regulation of the ecosystem?

―――――

If we look at population changes in the very short term, the changes from year to year, we see a complex and interesting pattern related to disturbances in the environment. We have already noticed (chapter 3) that deaths of both young buffalo and wildebeest (those under a year old) fluctuated wildly from year to year, and these changes were not connected to the population size—that is, they were not density dependent. Instead, they disturbed the population (figure 8.1). When we looked at other species like topi, kongoni, and even elephant, we also saw big changes in the survival of yearlings from year to year. Not only that, but these changes in survival and hence numbers seemed to be similar in different species. When kongoni survival declined in the drought, so did that in many other species. What was producing these disturbances, and why were they often similar in different species? How did they affect the regulation and stability of the ecosystem?

Kristine Metzger,[2] who was my data manager and analyst, and John Fryxell, in looking at the underlying causes of the fluctuations in the Serengeti ecosystem, focused on one particular global phenomenon. The temperature on the ocean surface north of Australia influences the direction of ocean currents, which in turn affect climate around the world. The temperature has been recorded as variations from an average

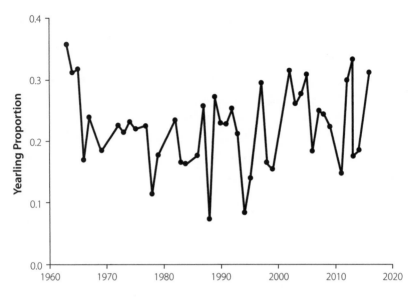

FIGURE 8.1. The ratio of yearling wildebeest (12–18 months old) to adult females is an indicator of calf survival, shown here from 1963 to 2016. The early high values were during the period when the population was increasing rapidly after rinderpest disappeared. Otherwise, the ratio fluctuated markedly from year to year. The question was what caused these variations. (Unpublished data.)

ocean-surface value by the Australian government since the 1880s,[3] and these variations above (positive) and below (negative) the average are known as the El Niño Southern Oscillation (technically called the *Southern Oscillation Index*). This affects weather cells around the world,[4] and eventually those effects reach the Serengeti. Their most important effect is on rainfall.[5]

El Niño affects dry-season rainfall, which in turn affects other parts of the ecosystem. A positive El Niño in the dry season produces more rain, more grass growth, and thus more grass food, which leads to higher survival of baby wildebeest (figure 8.2), topi, and even elephant. Even lions have more food because of the greater number of young wildebeest, which are easier to catch.[6] There are also more anthrax-disease outbreaks because more rain helps to spread the spores.

In contrast, when El Niño is negative there is less rain in the dry season but more rain in the short rainy season. This leads to more grass

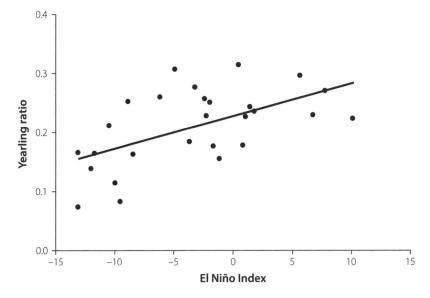

FIGURE 8.2. The survival of yearling wildebeest, as indicated by the ratio of yearlings to adult females, is related to the El Niño index. This index explains the variations in survival seen in figure 8.1. (Data from Sinclair, et al. 2013.)

growth during the short rains, resulting in more insects such as butter-flies,[7] and extreme outbreaks of rodents, which feast on the abundance of grass. Black-shouldered kites, black-headed herons, brown snake-eagles, and secretary birds, all of which depend on rodents to breed successfully, increase in numbers. Also, when El Niño is negative, total rainfall over the rainy season (October to June) is also higher, thus pro-ducing more grass and hence fuel for fires in the following dry season. Consequently, a negative El Niño in one year results in more burning in the following year. Overall, we see that different parts of the ecosystem respond depending on whether El Niño is negative or positive.[8]

The effects of El Niño were only detected after several decades of monitoring the different aspects of the ecosystem. El Niño explains in large part the causes of disturbances in many parts of the ecosystem—rainfall, fire regimes, plant growth, insect abundance, rodent outbreaks, bird breeding success, ungulate calf survival, lion survival, and even disease outbreaks. However, the two factors that are most affected by El

Niño are rainfall and fire, and disturbances to these have far-reaching consequences for the system.

———

Variations in rainfall, both over the landscape and year to year, strongly influence plants and animals, but particularly the migration of wildebeest (see chapter 6).[9] The long dry season lasts from July to October, but in the north and near Lake Victoria there are intermittent thunderstorms in September and October, which rarely occur on the plains. The lowest rainfall month is July, when fires are also at their most extensive, but since there is still some moisture in the soil from the previous rains, grass can sometimes regrow after fires. This regrowth, sometimes called "green flush," is only a few centimeters tall, the preferred height for wildebeest. The dry season coincides with the coolest time of year, the southern winter. Strong easterly winds prevail, and these contribute both to drying out the grass and to driving fires across the Serengeti. Later in the dry season the soil dries, and green flush occurs only after thunderstorms. There is also variation in when the rains start, sometimes early in October, sometimes as late as February, and sometimes they fail altogether.

The migrants respond to these variations (chapter 1). In the dry season wildebeest follow the thunderstorms around northern Serengeti, moving about 10 kilometers per day in small groups. When the rains start, there is a general movement south, determined entirely by the timing of the rains. Because the plains have much lower rainfall, the rains there normally start later, so the migrants usually wait at the edge of the woodlands until rain appears.

Rain affects all components of the ecosystem. Years with heavy rain produce tall, rank grass which provides fuel for fires later. Tree seedlings germinate in such years to form dense stands of similar age later. Insect numbers increase with the tall grass, and with extreme heavy rains come outbreaks of grasshoppers, beetles, butterflies, and moths. The breeding seasons of birds are tied to the rains, occurring on average a month after the peak of the rains in seed-eating and insect-eating birds, and two months after the peak in birds of prey, because they feed on the smaller birds and rodents.[10] Populations of birds that live in the long grass, such as guineafowl and white-browed coucal, fluctuate with the amount of rain.[11] Migrant bird species arrive in November, mostly from Asia but a few from southern Africa. This coincides with the first storms of the season, which are patchy. These migrants are also nomadic, following the storms around because storms bring insects (nearly all species are insect eaters). Some of these insects are migrants themselves, such as the armyworm moth and the locusts, but many others emerge from their hibernation (*aestivation*, in the jargon) with the first storms in huge numbers—ants, termites, beetles, and moths.[12] Rodent numbers show outbreaks in high-rainfall years; this affects rodent-eating raptors. Snakes also respond to the rodent outbreaks, and many more puff adders, spitting cobras, and mambas are seen.

———

Fire is an environmental factor that has a major disturbing influence on the ecosystem. It has been well studied over the decades by Mike Norton-Griffiths and others and synthesized elegantly by Stephanie Eby, of Syracuse University.[13] Weather fluctuations have major impacts on fire, and consequently vegetation and animal populations. Fires in the Serengeti occur largely in July, when the grass is dry, very long, and provides the greatest amount of fuel for burning. They are almost entirely caused by humans, and have been for millennia (see chapter 1). In July hunters, both past and present, set fires to attract ungulates to the green flush. In recent years these activities have been added to by park rangers who set fires because it is easier to chase hunters, and tourists set fires to get around. Fires caused by lightning are rarely recorded because lightning occurs during the September–October thunderstorms, when all areas have already been burned or grazed, so there is no fuel to burn. In years with high wet-season rainfall and low dry-season rainfall, plentiful grass followed by rapid drying results in extensive, hot fires.[14] Fires are far more common in the higher rainfall areas because of the taller grass; there is a direct relationship between the amount of rain that falls and the area that is burned in the next dry season in northern Serengeti.[15] Fires are almost absent on the short-grass plains because there is nothing to burn. Similarly, there are almost no fires outside both the western and eastern boundaries of the protected area because domestic animals eat up all the grass, and also because since 2010 illegal grazing has encroached along the whole eastern boundary, reducing fire events and radically altering the vegetation.[16]

Grazing by migrants has a strong influence on fires. Immediately following the rinderpest outbreak in 1890, fire incidence declined because people abandoned their farms, and it is people that light fires.[17] In the years 1920–70, heavy burning was practiced across the entire Serengeti ecosystem and west of it to prevent tsetse fly (which kill cattle and people), causing a drastic drop in the number of trees in Serengeti.[18] As the wildebeest population increased through the 1960s and '70s, their grazing reduced the amount of grass, and the area burned declined as a consequence (for the close relationship between wildebeest numbers and area burned, see figure 5.6).[19] As a result, trees regenerated, producing dense stands of nearly all savanna tree species by the year 2000 (see figure 5.7).[20] Grazing may also be involved in how soon a fire burns after the last one, called the *fire return interval*. Grasslands have the longest interval, nearly three years, compared to half that in dense savanna thickets; this is because wildebeest prefer to graze open grasslands and avoid dense stands, which are dangerous, as shown by Grant Hopcraft, and have lots of bothersome tsetse flies, so fuel for fires is much higher in the thickets.[21]

Fire has destructive effects at many levels. It reduces nitrogen and detritus in soil and so lowers soil productivity.[22] Fires burn back small savanna trees to their roots and keep them from developing to maturity, as demonstrated experimentally by Holly Dublin, and they reduce regeneration of riverine forests, as shown by Greg Sharam[23] (figure 8.3). Ally Nkwabi, one of our senior researchers, conducted some elegant experiments in the 2000s using fire and grazing to show that fires reduce the abundance of insects[24] and destroy the nests of small birds that nest in long grass.[25] Buffalo avoid burned areas because the grass height is too low for their mouths, and lions avoid burned areas probably because there is insufficient cover for successful hunting.[26]

FIGURE 8.3. Since the 1960s, human-set grass fires have been destroying Mara riverine forests because these fires are able to penetrate past the forest edge. Kogatende, northern Serengeti, August 1998.

Yet fire has many positive effects too. Smaller antelopes such as oribi, Thomson's gazelle, Grant's gazelle, impala, warthog, and even wildebeest are attracted to the green flush after burns.[27] In general, the smaller herbivores are attracted to burned areas, while the larger ones are not. The reasons could be that burned areas produce young shoots high in nitrogen, the preferred food of small but not large herbivores,[28] and these areas are also safer habitats from predators due to decreased vegetation height and increased sighting distances.[29]

When fire reduces long-grass habitats to short grass, and green-flush sprouts, short-grass bird communities from the eastern plains move into these burned areas, at least temporarily. Foxy larks (fawn-colored larks) and Athi short-toed larks move out, while red-capped larks and capped wheatears move in.[30] Dusky larks are adapted to feeding only in recently burned areas, and crowned plovers nest in the green-flush regrowth.

Both these species are nomadic and follow the fires. Immediately in front of active fires huge swarms of insects fly out to escape, and these are followed by birds such as lilac-breasted rollers and various shrikes and hornbills. Snakes and rodents also run before the fire and are eaten by tawny eagles, bateleur eagles, and black-shouldered kites. Immediately behind the fire are flocks of white storks, Abdim's storks, and superb starlings eating the recently killed insects.

————

Changes in rain and fire regimes create the disturbances that produce changes in many aspects of the ecosystem, including fluctuations in animal numbers, and it is these fluctuations that hide the equilibrium. We see that El Niño provides the patterns of rainfall and fire that underlie the disturbances to all aspects of the ecosystem. What started as simple descriptions of these fluctuations in Serengeti turned out, when we looked closer, to be highly complex processes. All ecosystems are disturbed by changes in the environment. But regulation counteracts these changes and keeps the system from drifting to extinction, and so maintains persistence. This was clearly demonstrated by two almost perfect (if unintentional) experiments: when drought knocked down the wildebeest in 1993 and numbers subsequently returned to their previous level, and earlier when poachers knocked down buffalo and elephant numbers in 1978 and both also returned to their original levels, all due to regulation. Disturbance and regulation work together.

9

Continuous Change in Ecosystems

In July 1965, I looked out over the plains from one of the hills at the edge of the woodlands near Seronera and was overawed by the feeling of timelessness in the pristine expanse of endless grassland, parkland, hills, and rivers. I felt a sense of permanence; this savanna had been here, it seemed to me, since time began. And this sense was reinforced by the knowledge that Serengeti was legally protected inside a boundary forevermore. Serengeti would be here for everyone, a special area, one of the great wonders of the modern natural world. It was one of the last remaining places that resembled the world as it had been before the end of the Ice Ages, during the Pleistocene, more than 10,000 years ago. During that period there were many migrations of large mammals all over the world, but most are now gone, the remaining ones highly modified. Serengeti was outstanding, one of the last examples of a time long gone.

———

As my work proceeded over the decades, monitoring and measuring the ecosystem, I gradually became aware that this idyllic view of permanent, pristine nature may not be true. I found things changing in just about every aspect that I looked at. I was forced to consider that these changes were either unusual, aberrant conditions in present time, or they reflected the true behavior of the ecosystem over long time periods. It was this realization in 1980 that spurred me and my colleagues to start a program of measuring change, particularly in grasslands, savanna trees, forests, fire regimes, herbivore reproduction and population size, and predators.

I had been working with Mary Leakey since 1973. She had made her name, along with her husband, Louis Leakey, excavating the hominid remains at Olduvai Gorge since the 1930s, and she knew a lot about the prehistory and paleontology of the Serengeti. She had just discovered the footprints of an early man that had lived some 3 million years ago.[1] Mary, Mike Norton-Griffiths, and I had worked on migrating ecosystems in the 1970s,[2] and it was this collaboration that made me aware of the great changes that had taken place in our ecosystem over millions of years. After Mary retired and her work was taken over by Robert Blumenschine and colleagues at Rutgers University, the work eventually produced the vitally important story of change in the Serengeti. They showed that the plains have been changing continuously in their ecology and habitats for 4 million years.[3] Their studies focused on the excavations at Olduvai Gorge, which cuts across the Serengeti plains, and at Laetoli, a site on the edge of the plains at the foot of the Crater Highlands. Four million years ago, this area supported dense woodland; gradually the environment became drier and habitats changed progressively to bushland (2–3 million years ago), open savanna (1 million years ago), and finally (in the last 500,000 years) to the present semiarid short grasslands (figure 9.1). There is every expectation that habitat change will continue irrespective of human-caused climate change.

These vegetation changes imply a gradual drying trend partly due to the development of the Crater Highlands, which create a rain shadow, and partly due to long-term trends in climate. We know that over the past 135,000 years the climate has been highly variable, based on sedimentary cores from Lake Malawi, a large, deep African Rift Valley lake south of Lake Victoria.[4] Based on lake levels one can see two very dry periods lasting 8,000 years (135,000 years ago) and 20,000 years (115,000–95,000 years ago), respectively. The lake dropped some 600 meters during each of these periods. This level of drop means that Lake Victoria would have been completely dry, and aridity would have been extreme, more severe than any period in the past 100,000 years. Precipi-

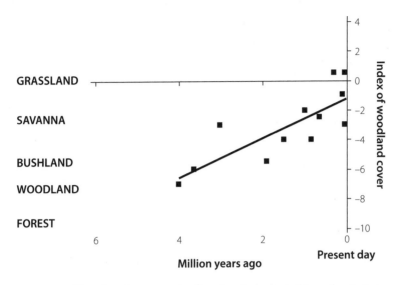

FIGURE 9.1. The index of tree vegetation (based on the content of the carbon isotope ^{13}C in the soil layers at different time periods in the past) shows a progressive change from dense woodland 4 million years ago to open treeless plains at present. Change has been continuous. (Data from Peters, et al. 2008.)

tation was probably less than 400 millimeters per year, and a cool semi-desert would have existed in the Serengeti region.[5] The cool, dry conditions during the arid periods were favorable for the spread of montane forest containing trees of *Podocarpus, Olea, Juniperus,* and *Ilex* around Lake Malawi, all species that are found in the Serengeti montane forests today. The climate became wetter at about 95,000 years ago and reached near-modern conditions about 60,000 years ago. Fire frequency returned to something similar to the pre-rinderpest conditions of the nineteenth century.

The last glaciation, 35,000–15,000 years ago, was a period of extreme drought throughout Africa. We know this from lake cores in both Lake Malawi[6] and Lake Victoria.[7] Conditions were 4°–7°C cooler than today.[8] Lake Victoria dried out during this period and only started to refill 12,000 years ago. There was no lake-derived rainfall, which accounts for about 50 percent of present-day rain. However, during the last glaciation, periods of drought alternated with wetter conditions lasting several thousand years.[9]

Following the end of the Ice Age approximately 12,000 years ago, the climate warmed, although there were also rapid shifts. There were warm, wet periods when rainfall increased to possibly double that of today and lakes rose some 100 meters above present levels. In the Gregory Rift, just east of Serengeti, a paleo-lake covered the whole Natron-Magadi basin to a depth of 50 meters. Lake Victoria rose 100 meters and flooded east across the Ndabaka floodplain for some 50 kilometers.[10] During this humid period there were three sharp shifts to drought conditions.[11] The drought at 4,000 years ago was severe, lasting 300 years.[12] It caused the collapse of the Egyptian Old Kingdom. The Nile River nearly stopped flowing, which suggests that Lake Victoria was close to empty.

Over the last 2,000 years, there have been several fluctuations in climate. In the Medieval Warm Period (1000–1270 CE), conditions were dry,[13] but in the following Little Ice Age (1270–1850 CE) conditions were cool and wet.[14] Strangely, this is the exact opposite of what happened in the more distant Ice Ages, when cool conditions resulted in drought and Lake Victoria dried up completely. Over the past 200 years we see that climate was drier in the early 1800s than subsequently but then became very wet during 1890–1910.[15]

For the last 100 years, rainfall has gradually increased in the Serengeti area, based on records at Musoma and Shirati from 1902 to 2009 on the east side of Lake Victoria, and at Banagi, in the center of the Serengeti.[16] The wettest year, 1961–62, raised the lake level several meters. Temperatures over the African continent have warmed at a rate of 0.5°C per century.[17] Over the past few decades in Serengeti (1960–2007), maximum daily temperatures have been increasing. However, rainfall is strongly dependent on Pacific Ocean surface temperatures, which show no long-term trend. Instead, rainfall shows multidecadal oscillations with the past few decades becoming drier.[18]

These studies of ancient climates brought it home to us scientists working in present times that the Serengeti was far from constant; it has experienced both long and short periods of drought and wet from several thousand years to a few hundred years to a few decades in duration. It might appear that there were long periods of drought or flood in the distant past compared to shorter periods recently, but this is an artifact of our ability to detect short-duration changes from the geological records in distant times. If the Serengeti's climate is constantly changing, this suggests that the ecosystem must also be constantly changing. I wondered, in 1980, how the ecosystem responded to climate change and whether I could detect such changes.

———

Over long periods of time, species come and go in any community. Such slow change in communities has been most clearly recorded in plants after the Ice Age of North America. For example, as conditions became warmer over time scales of hundreds or thousands of years, various tree species spread north at different rates, so that at any one place the community was an ever-changing combination of species.[19]

In East Africa there is evidence of alternating plant communities driven by changing climate going back 1,000 years or more. Pollen records at Tsavo National Park in Kenya show an alternation between savanna and grassland.[20] Savanna and dense bush has been present in Tsavo since 1700, but before that the area was open grassland for 400 years, and before that, savanna again. In the alternation between two different habitats, the transition is often very quick, just a few decades. Can we see changes in Serengeti vegetation that are due to this constantly fluctuating climate? During the Medieval Warm Period (950–1250), when conditions were, as the name suggests, warm and dry, the montane forests would have retreated to refuges on hilltops on the eastern border of Serengeti. It is likely that the lowland tropical forest spread east from Lake Victoria up the Grumeti and Mbalageti rivers. Conversely, in the Little Ice Age (1270–1850), cool and wet conditions allowed the expansion of montane forests west along the Mara River and across the whole area north of the northern Grumeti River.

In looking for evidence of these supposed changes in our plant communities, I found a few clues. First, small patches of montane thickets are today spread over the whole of the northern Serengeti and the Mara Reserve. Montane forest still occurs on top of the highest hills in the northeast of the ecosystem. These remnant patches indicate that at one time in the not-too-distant past, contiguous montane forest spread across the whole of northern Serengeti. When would this have been? It is likely that this forest could have spread during the Little Ice Age, especially in the early 1800s. We also know from historical records[21] that agricultural tribes had begun an expansion after 1850—these would have been west of Serengeti—and that the pastoral Maasai, who previously had not reached Serengeti, had also arrived around 1850.[22] All this suggests that montane-forest expansion in northern Serengeti took place about two or three centuries ago and has since been gradually reduced by burning practiced by expanding human populations. Holly Dublin, one of our students in the 1980s, and Greg Sharam in the 2000s[23] have both found that these forests and thickets are now being reduced largely by intentional burning, which has increased over the past century. The montane forests, as we had suspected, may well have stretched along the Mara River all the way to Lake Victoria.

Lowland forests may have had a different history. They would have spread from their refuges around Lake Victoria and the Congo forest up the Grumeti and Mbalageti rivers during warm periods. Greg Sharam has found that in Serengeti these forests can only regenerate when they are protected by nursery trees, the very thorny white river thorn, *Senegalia (=Acacia) polyacantha*.[24] We have seen this protection and regeneration take place over the last 40 years, with a massive thicket of young *S. polyacantha* that started on the outer edge of the Grumeti forests in 1978. These were huge trees, some 25 meters tall a mere 20 years later, an exceedingly fast growth (figure 9.2). Then, in about 1998, inside these stands young forest trees began to grow, and by 2020 they had reached half their mature height at about 10 meters. No forest has developed outside these *S. polyacantha* stands. A careful look inside the current mature forest along the rivers shows the occasional very large *S. polyacantha* amid tall forest trees. Since *S. polyacantha* cannot regenerate inside forest, these sentinels show that at one time the present forest was absent and began, as did the others, within the protection of *S. polyacantha* stands. African acacia are relatively short-lived species, and our *S. polyacantha* stands are already showing signs of senescence and death. It is unlikely that they live longer than about

FIGURE 9.2. *Top*: Riverine grassland to the right of the curve of riverine forest on the Orangi River, Kimerishi Hill western corridor, June 1980. *Bottom*: The same area 20 years later, in May 2000, showing the stand of riverine trees (*Senegalia polyacantha*), now 20 m tall, that has taken over the grassland. Note also the increase in density of savanna trees on the hill slope.

150 years, which means our forests are only about that age. They could have started after the Little Ice Age, about 1850, when the wet climate provided favorable conditions for the forest to expand.

Another clue to the history of the lowland forest comes from their current distribution. On the Grumeti River they are found at the lower end, near Kirawira Guard Post, some 35 kilometers from Lake Victoria. This is the point that the lake would have reached during the wet period of the Little Ice Age (1270–1850). Forest does not grow in the seasonally inundated floodplains that are currently present, and it seems it has never spread downstream despite the retreat of the lake. The forest grows for some 70 kilometers upstream on the Grumeti, varying in width from a few to several hundred meters. But they stop at Kimerishi Hill, just west of Banagi. Upstream of this there is only a fringe of green acacia (*Vachellia kirkii*) along the banks. But at one time this forest must have extended another 50 kilometers, as far as the eastern park boundary. I know this because one of the characteristic subcanopy trees, *Cordia goetzei*, which is found only in closed-canopy lowland forest, occurs occasionally on top of isolated termite mounds in a thicket of small bushes near the river at Klein's Camp. Clearly, lowland forest extended this far and then retreated, leaving minute remnants that are now fast disappearing. These remnants are a few hundred meters from the montane thickets on the slopes of Kuka and Lobo Hills, indicating that at one time the two forests may have been close together.

A further clue that supports this interpretation involves black-and-white colobus monkeys. There are generally two subspecies of this monkey in East Africa, a lowland form and a montane type. In Serengeti there is a rare highland subspecies, which occurs in the montane areas on the northeastern boundary near Lobo. But this rare montane form also occurs in lowland forest on the Grumeti and Mbalageti rivers— the subspecies has become adapted to a completely different forest. There is now a gap of some 70 kilometers between the lowland and montane forests at Lobo. I suspect that the monkeys got to the lowland forests when these extended eastward to the hills two centuries ago, and then got trapped when the forest retreated.

In the 1960s conservationists, scientists, and park wardens became aware that savanna trees were fast disappearing all over Africa. Most of them blamed elephants, whose numbers were increasing after ivory exploitation ceased in the 1880s (see chapter 4) and who were now feeding on the abundant savanna trees, damaging and destroying them. Wardens instituted culling to control elephant numbers and maintain the trees in the abundance they assumed represented the "real" Africa, the Africa they had seen when they first arrived, their baseline of reference, discussed earlier in this chapter. Thousands of elephants were killed in Murchison Falls National Park, Uganda; in Hwange National Park, Zimbabwe; and in Kruger National Park, South Africa.[25] Tsavo National Park, Kenya, was also losing trees, and scientists there recommended culling, but in this case the wardens refused.[26] In Serengeti the increasing numbers of elephants were also causing concern, and by 1967 culling was discussed.[27] And then Mike Norton-Griffiths arrived, in 1968. As mentioned earlier, he was unimpressed with what seemed to him simplistic arguments by scientists and wardens alike. What he took note of, instead, was the amount of burning taking place: a remarkable 90 percent of the park was set alight intentionally by hunters and managers in the dry season (as described earlier). He set about monitoring the burning each year, and from aerial photographs taken in 1953, 1959, and 1971 he measured the numbers of large trees. What he found was, as elsewhere, a declining number of trees but also, because of the fires, a complete absence of young trees. Most importantly, he showed that the rate of loss of large trees caused by elephants was trivial compared to the loss of young trees caused by fires. He developed computer models clearly showing that reducing the area of burning would allow a huge regeneration of young trees, easily outstriping the loss from elephants. In a hectare there are thousands of baby trees, while only a few large trees are killed by elephants. It was obvious that it was fire, not elephants, that was causing the disappearance of trees.[28]

Even more telling, however, I found that the loss of large trees had been taking place continuously since the 1920s, well before elephants had arrived in Serengeti in the 1950s. I knew this from a chance discovery. Although scientists had only begun work in Serengeti in the 1960s, wardens, tourists, and hunters had been there since the 1920s, and they had taken photographs. In 1980 I began looking for all the photos I could find that had identifiable backgrounds that I could use to locate those places again. Myles Turner, the longtime warden of Serengeti in the 1950s, had a few photos. Photographs taken in 1944, by the professional hunter Syd Downey, of the western side of the Mara Reserve in Kenya showed extensive woodlands of acacia trees. This was part of the northern Serengeti ecosystem. When I repeated these photographs in 1984, 40 years later, only two trees remained in the same area.[29] We will return to this example in chapter 10 (see figure 10.1).

By far the most important contribution of early photos came from the filmmaker-hunters Martin and Osa Johnson.[30] On their second expedition to East Africa, they spent six weeks in Serengeti in July 1926. The vegetation was dense, mature *Commiphora* trees and young regenerating umbrella acacias, very different from the same area in the 1980s, when it was open acacia parkland similar to that in the 1890s described by Baumann (see chapter 1).[31] On their third expedition, they spent much of May to July 1928 camped at Moru and traveling along the Mbalageti River. Their fourth expedition was a grand affair with two amphibious biplanes, which they used to make aerial films of the wildebeest migration.[32] During all these trips they took extensive photos, mainly of lions, since these were the object of their films, but in the process they photographed the vegetation and the skyline, which I could use to find the places again. In some of these photos I could find the exact same tree again, 60 years later. Figure 9.3 (*top*) is an example of a Johnson photo taken near Lake Magadi in southern Serengeti in 1928. I repeated the photo in 1982 and again in 2003, three-quarters of a century after the first one (figure 9.3, *middle and bottom*). The original *Commiphora* woodland had disappeared by the 1980s, but there was subsequent resprouting of different trees, including the umbrella acacia (*Vachellia tortilis*). From this bonanza of photographic sources, I could calculate the number of trees and see how this number had declined from 1926 to 1980 (figure 9.4).

FIGURE 9.3. *Top*: View of the Nyaraboro escarpment from Emakat Hill at Lake Magadi, showing stands of *Commiphora* trees in the foreground. Photograph by Osa Johnson in 1928 (with permission from the Martin and Osa Johnson Safari Museum, Chanute, Kansas). *Middle*: The same view 54 years later, in May 1982, showing the disappearance of *Commiphora* savanna, replaced by grassland. *Bottom*: The same view 20 years after that, in June 2003, showing the reappearance of savanna, this time dominated by umbrella trees, *V. tortilis*.

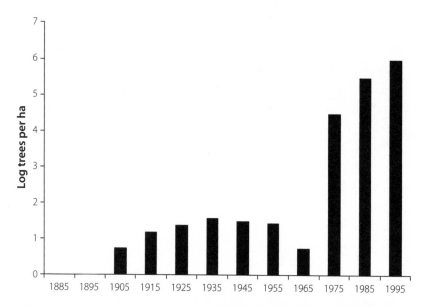

FIGURE 9.4. The abundance of trees that began growth in different decades since the 1880s.
One cohort of trees appeared in the 1890s and dominated the first half of the
20th century, declining in the 1960s and early 1970s due to severe burning.
A second cohort arose in the 1970s due to wildebeest grazing and is still progressing.
(Data from Sinclair 1995 and unpublished data.)

What was the reason for this decline in trees? After rinderpest had
caused the evacuation of people surrounding the Serengeti ecosystem
(see chapter 2) in the late nineteenth to early twentieth century, dense
thickets of trees and shrubs had grown up, bringing with them the tsetse
fly and sleeping-sickness disease.[33] This was the dense vegetation that
can be seen in the early Johnson photos; it was present throughout the
1900–1920 period during most of the German administration (1890–
1921). In 1922 the British government took over, and one of their policies
was to reduce the impact of the tsetse fly on people and cattle. They did
this by instituting a program of intense burning to open and clear the
vegetation.[34] The policy of burning became the standard dry-season
activity over the next 50 years. It had the desired effect of reducing the
tree population, and concomitantly the tsetse fly.

Then, in the 1970s, fire frequency declined because of grazing by the
increasing numbers of wildebeest, now recovering from rinderpest.

Wildebeest ate up the grass before fires could take hold, and the area burned shrank as the wildebeest population grew (figure 5.6).[35] In 1980 I set up a series of photographic points on the tops of hills from which to record the tree populations below. My intent was to document the change in tree numbers as they declined from the fires. I had not yet appreciated the impact that wildebeest were to have on the fires—and the trees. I anticipated that all the trees would eventually disappear, leaving an open grassland. Six years later, when I came to photograph those sites again, I found instead a mass of small regenerating acacias, which by the 2000s had developed into big trees (figure 9.5). In fact, all over the ecosystem inside Serengeti there was a massive regrowth of baby trees of all savanna species including acacias and broad-leaved *Terminalia* and *Combretum*. This was exactly what had been predicted by Mike's analysis of the fires.

I had another lucky clue that confirmed these observations. I knew the exact age of some of the main trees—the umbrella acacia, stink-bark acacia, and Kirk's acacia—because they germinated in 1970, when the Serengeti Research Institute was built, and I have measured their diameters at intervals for almost 50 years. In addition, I have been able to use some of the Johnson photos to measure the tree diameters from the 1920s. From this I was able to get a rough estimate of the age of the acacia trees in Serengeti. I found that there were two major age groups: one younger cohort appeared in about 1980, after wildebeest reached their peak and burning was much reduced because of herbivore grazing; and an older cohort that is now dying of old age but started in about 1890, after rinderpest hit and drove out both the agricultural and pastoral tribes and so reduced burning too (figure 9.4). The important point is that savanna is always changing, not in smooth transitions but in sudden pulses.[36]

———

Apart from the black-and-white colobus, mentioned earlier, is there evidence for any other historical changes in animal populations? One can only surmise what may have happened as the forests expanded and retreated. When forests extended down along the Mara River, they would

FIGURE 9.5. View from Kimerishi Hill facing west along the corridor that shows the regeneration of savanna with stink-bark acacia (*V. robusta*) in foreground and wait-a-bit acacia (*V. mellifera*) in middle ground. *Top left*: 1980; *bottom left*: 1986; *top right*: 1991; *bottom right*: 2011.

FIGURE 9.5. (*continued*)

have been as wide as 10 kilometers (based on the location of hill thickets today) and quite impenetrable to migrant plains animals. Some animals may have found access to the river or its tributaries, but they would not have moved north into what is now Kenya. It is likely they moved farther west, to Lake Victoria. Indeed, Baumann found wildebeest drinking at Speke Gulf in 1891,[37] and the hunter S. E. White found wildebeest near the newly founded town of Musoma in 1913,[38] proving that the migrants were clearly aware that the lake existed and provided water. White also recorded that beautiful antelope, the greater kudu, an inhabitant of forest and thicket, in the Kuka Hills in 1910, but it has never been seen since. Giant forest hogs live in montane forest and still occur along the Mara River in Kenya. The warden Myles Turner saw a giant forest hog in 1972 in the riverine forest of the Mara River, seven kilometers west of Kogatende. He assumed it had wandered downstream along the Mara from Kenya. The previous presence of both of these species implies that forests were more extensive in the nineteenth century. Elephants were in Serengeti in the first half of the nineteenth century[39] but then vanished as the ivory trade cleaned out Africa (see chapter 4), and they only returned in the 1950s. Both roan antelope and wild dogs were widespread in the Serengeti savanna during the period 1910–60 when wildebeest were at low numbers; after wildebeest increased, followed by lion and hyena numbers, both roan antelope and wild dogs became extinct in most of Serengeti.

Climate change over the past few decades has caused changes in the distribution of bird species in Tanzania, most of them moving westward toward higher rainfall. Consequently, several species have expanded into the Serengeti. For example, the Taita fiscal shrike, Pangani longclaw, white-bellied go-away-bird, and yellow-throated spurfowl have all moved into the Serengeti ecosystem from the east since 1970. The endemic bird *Karamoja apalis* spread into Serengeti after 1994 following the increase of whistling-thorn acacia due to the increase in wildebeest grazing.[40]

These movements of bird species track the rainfall gradient, moving toward wetter areas. This implies that species already in wetter areas will move farther west, out of the park, if Serengeti becomes drier in the future, so as to stay in a region with the same rainfall. In fact, such movements toward wetter areas by birds in East Africa have now been seen: they can enter wetter protected areas such as national parks but cannot leave because they would be moving into agricultural land (west of the park), where habitat does not exist. This is a concern for conservation. Movement of these species is part of the more general spread of bird species westward due to human-caused climate change, which has been documented by Colin Beale.[41] Such examples illustrate how change in community structure resulting from environmental change can affect the regulation of the entire ecosystem.

———

If we look back over the historical record, we see that the environment of East Africa, and of Serengeti, has changed over periods of thousands, hundreds, and even tens of years, often very fast, and many times over. I have seen this for myself over just the past 50 years in Serengeti. Alternating periods of wet and dry, cool and warm conditions show that there has been no prolonged period of constancy in climate. The plant communities change with the environment, and animal populations react to the change in habitats. When Lake Victoria dried out, there would have been only a weak rainfall gradient and hence less motivation for a migration, while high levels in Lake Victoria would have produced the opposite effect. The migration thus would have appeared and disappeared many times, with the attendant "downstream" consequences for the ecosystem.

Now I think back to 1965, when I gazed across the pristine umbrella acacias, the distant expanse of plains, and the rivers with their fringe of fig trees and forest. A carpet of distant grazers, seemingly stationary in the evening light, a bull elephant drinking down at the river while a matriarchal group marches in majestic slow motion across the plain. The immutable scene of peace—but it was all a glorious illusion. The Mara River forests have come and now largely disappeared, the savannas have changed in the past 100 years, now much denser than they were but, more importantly, much changed in species. The plains have gone from long grass to short as the migration has moved out to the east. The dominant ungulate was once the Thomson's gazelle but is now the wildebeest, and elephants were not even in Serengeti 100 years ago.

I realized that to find the causes of these changes I had to look back in time, not just to the recent past but to prehistory, because events during those ancient times set the scene for the present: they determined the geology, geography, and evolution of our present environment and provided the clues as to why changes are now taking place. By great good luck, the Serengeti has been the site of considerable paleontological research, which has given us an understanding of what the environment and species were like from millions of years ago up to present times.

The sense of permanence we experience when we first see a place is misleading, and it also has important conservation consequences, which we will explore in chapters 11 and 12. The truth is that *all systems change all the time*. Nothing stays the same. Regulation, therefore, is tracking a moving baseline, a moving equilibrium.

10

Appearance of Multiple States and Rapid Shifts in Ecosystems

Things were changing in the Serengeti; that is the conclusion I reached in the last chapter. But there was one serious problem that told me that I did not know the whole story. Things were changing in one place but not another. How could this be?

As described in chapter 9, the number of trees declined from the 1920s to the 1960s because of increased burning. They started to recover from the 1970s onward due to the increasing wildebeest population, which kept the grass short and prevented fires from spreading. However, there was one outstanding and inconvenient problem for this theory of burning: whereas in the Serengeti National Park the reduction of burning recorded in the 1970s resulted in the appearance of a mass of young trees, no such response was occurring in the Mara Reserve of Kenya immediately adjacent to and connected with the Serengeti. There was no natural boundary between the two areas; it was just a line on the map. Wildebeest grazing was just as heavy on the Kenyan side, probably even more so because this was their dry-season refuge. So why was there so little tree regeneration in the Mara Reserve—in fact, none at all?

We knew from the photos taken by the professional hunter Syd Downey that Mara had once been home to extensive savanna. Downey had used the Mara area of Kenya and the Lamai of Serengeti almost exclusively as his hunting grounds in the 1930s and '40s; one of his hunting blinds, built in a tree, remained there until 1987, when the tree died and collapsed.[1] He took clients hunting in this area—it was wild, remote, and teeming with wildlife, especially with rhino, buffalo, and lions, as well as with wildebeest in the dry season. Elephants were present but not in large numbers. A photograph taken by Downey in 1944 from one of the characteristic small hillocks on the western side of the Mara River, in what is called the Mara Triangle, shows a landscape with so many trees that it is difficult to see the grass. A photo taken in 1984 from the same point shows only two trees remaining in a sea of grassland, and the same area photographed by Reto Bühler (courtesy of Joseph Ogutu) in 2018 shows that grassland continues to this day, maintained by elephant browsing (figure 10.1)[2]. Clearly, trees were quite capable of growing in the Mara grasslands, but they were not growing there in 1984 when in Serengeti they were growing everywhere. The open grassland we found in Mara in the mid-1980s should have shown the same regeneration that we were seeing in Serengeti. Evidently, there was more to this story than we had understood so far. This was the inconsistency that told me that something else was going on—and it is inconsistencies that are the best clues for scientists.

I decided to set up a series of experiments to find out why trees were not regenerating in the Mara Reserve, when they were doing so in Serengeti, and gave the task to one of my students, Holly Dublin.[3] Holly did something others had not done in the Mara Reserve—watched elephants feeding. By careful documentation of what was destroying small trees, combined with experiments on burning and the exclusion of antelope browsers, Holly reached a surprising result. Mike Norton-Griffiths had already shown, for the Serengeti, that fires killed small trees so that they could not replace large trees, and that the large trees had usually died of old age (mainly in the 1950s and '60s) and not from elephant damage. Holly confirmed this result for the Mara Reserve. However, she discovered a different and remarkable effect of elephants on tree populations: elephants could eat enough small seedlings, even when fires had effectively been prevented by wildebeest grazing, to prevent trees from returning and keep the vegetation as a treeless grassland.

FIGURE 10.1. *Top*: View in 1944 from hillocks in the Mara triangle showing dense savanna of *Vachellia gerrardii*. Photocopy of a photograph by the professional hunter Syd Downey of his wife, Cynthia. (Copy donated by Mrs. Downey and published in Dublin in 1991.) *Middle*: The same view 40 years later, in September 1984, showing the complete change from savanna to grassland. *Bottom*: The same hillocks after another 34 years from a lower viewpoint, showing that grassland continues, maintained largely by elephant browsing of seedlings. Compare with regeneration in Serengeti during the same period (see figure 9.5, *bottom right*), 2018. (Photograph taken by and used with permission of Reto Bühler, Zurich, Switzerland, with the help of Joseph Ogutu, Stuttgart, Germany.)

FIGURE 10.2. If sufficiently numerous, elephants are capable of pulling out most small acacia seedlings and so maintain open grassland, as seen in the Mara triangle of Kenya, August 1984.

In other words, when fire had been precluded by wildebeest grazing, tree saplings still did not return because of elephant browsing. Another factor had entered the picture (figure 10.2).

However, this was still only half the answer. Why was the converse not true? If wildebeest and antelope grazing, and elephants, were required to prevent regeneration in the Mara Reserve, why was this not also happening in the Serengeti? Since grazing was the same on both sides of the border, the great regeneration of trees in Serengeti would suggest one thing: an absence of elephants. And this was indeed the case in the 1970s and '80s.

We had been counting elephants throughout the Serengeti ecosystem since 1965 (see figure 4.6). Initially numbers increased, reaching a peak in the mid-1970s. Our count of 1986 came as a surprise, then, because it showed that elephant numbers in Serengeti south of the Kenyan border had been reduced to only 10 percent of their number in the 1970s. In contrast, elephant numbers in the Mara Reserve had not changed—if anything they had increased a little. The border was the dividing line because there was a significant difference in antipoaching efforts on either side. In the Mara Reserve there was almost no poaching because of Kenya's efficient antipoaching field force. In Serengeti there was effectively no antipoaching effort, and elephants were being killed at a high rate, targeted for their ivory. Elephant numbers had been at around 3,000 in 1976, but by 1986 there were a mere 400, and these animals were highly traumatized, running at the first disturbance and the sight of a vehicle.[4] Most were cowering on the short-grass plains near Lake Lagarja, a habitat completely unsuitable for them as a source of food but where at least they felt safe—ivory poachers did not go there. Ivory poaching was rampant from 1978 to 1986, especially 1982–84, when one could hear rifle shots every night, and carcasses were found lying by the roads the next day with their ivory cut out within sight of park headquarters. Thus all indicators pointed to poaching as the reason for the demise of local elephants.[5]

In Serengeti, therefore, there was no elephant browsing, and small trees were able to regenerate. In the Mara, elephant browsing was still heavy and prevented tree regeneration. The results seemed to confirm the theory of Mike Norton-Griffiths and Holly Dublin that a combination of fire, wildebeest grazing, and elephant browsing was needed to explain the changes in vegetation, a far more complex interaction than we had imagined. Elephants can live and feed on trees without much reducing tree numbers; the same number of elephants can feed in grassland and prevent trees from regenerating. Consequently, we had the interesting situation of two *ecological states*, both with elephants but one with many trees and one with relatively few. And these two states occurred at the same time, one in Serengeti and one in Mara. Elephants cannot modify the system from one state to another, but they can hold it at the low state, preventing it from regenerating and changing into a high state. Put differently, elephants could not regulate high density of trees (bottom-up control) but could regulate low density of trees (top-down control).[6]

———

We saw in the last chapter that ecosystems are continuously changing, often due to the disturbances discussed in chapter 8. If the system is pushed too far, however, it will suddenly change into a new configuration of species and processes. When this happens, we say that the system is showing *multiple states*. It is easy to identify multiple states because we observe them this way: if a system returns to its original state once the cause of a disturbance has been removed, then the system has one state; but if the system does not return to its original state but stays in a new state on removal of a disturbance, then it shows multiple states—two and perhaps more.[7] In the case of the Serengeti trees, fire had reduced their number, but when fire was itself reduced, the trees returned. This is one state. However, in the Mara Reserve, fire had reduced the number of trees, but when fire was also removed, trees did not return—their numbers remained low because another factor, elephant browsing, had taken over. Two states: one with savanna and elephants, the other with grassland and elephants. Professor Buzz Holling, one of my colleagues at the University of British Columbia, was the first to propose the idea of multiple states in 1973.[8] In discussions with him I realized I could test this idea in Serengeti. This became one of the first examples of multiple states documented in nature.

———

We now know that there are several other examples of multiple states in Serengeti. John Fryxell, using both the data from his studies in Sudan and those of the Serengeti (chapter 6), constructed mathematical models describing the regulation of migrating and resident populations of ungulates, and he made an interesting discovery. If the migratory-prey numbers were very low, the predator population was able to regulate the prey population—in other words, keep numbers at some low level. If, however, prey numbers somehow increased, then predators could not keep numbers down, and the prey population climbed to a much higher level set by the food supply. The prey population could stabilize at either a low level limited by predators (top-down) or at a high level limited by food supply (bottom-up). The system, therefore, had two possible stable states.[9]

In 2013 we had the good fortune of discovering previously unknown data from ground counts on the Serengeti plains conducted in February 1953 by the first scientist in Serengeti, Allan Brooks, a Canadian sent to study Thomson's gazelle.[10] Although these data are crude by today's standards, they suggest that gazelle was by far the most numerous species, some 500,000, while wildebeest were only about 100,000, not much greater than rough estimates from 1933 deduced from aerial photos taken by the Johnsons. If the numbers can be believed, they suggest that the wildebeest population was stuck at a low level for many decades, kept there by a combination of rinderpest mortality of calves and predation. This may be a second case of two stable states produced by predators, as discovered by John Fryxell.

———

Predators can also have indirect effects that produce multiple states by influencing biodiversity and the interactions between species. Top predators, particularly the ones we call keystone species (described in chapter 7), maintain a diversity of prey species with many competitive and mutualistic interactions among them; this creates a stable system. If a disturbance alters the competition, species may be lost, and even if the disturbance is removed, the lost species may not return, thus changing the community to another state.

Greg Sharam, one of my students, whom we met earlier, found just this situation in the riverine forests along the Mara River in northern Serengeti. Greg showed that on the Mara River, riverine forest is renewed through seedlings growing up underneath a closed canopy.[11] These seedlings grow from seeds that have been previously eaten on the tree by fruit-eating birds such as hornbills, barbets, starlings, turacos, and bulbuls. The seed itself is regurgitated or passes through and is deposited on the ground without its outer fleshy cover (the pericarp). Such naked seeds have a high germination rate and provide adequate regeneration for the forest—it is a stable system. Birds are well-known dispersers of fruit and seeds in forests around the world and act in a keystone role.[12]

Disturbances caused by fires that burn into the forest from surrounding grasslands or by elephants browsing, if sufficiently severe, can open up the canopy (see figure 8.3). Then the fruit-eating birds disappear because they require a dense, closed-canopy forest. The seeds drop to the ground, but now still enclosed by the pericarp, and in this state they are attacked by bruchid beetles (tiny insects a few millimeters long). The beetles lay their eggs on the seeds, and the larvae burrow in so that the seed fails to germinate, or if it does, germinates at a low rate, around 15 percent, which is not sufficient to maintain the forest. The few that do germinate are killed by dense, rank grass that invades openings in the forest. Without its fruit-eating birds, the forest gradually dies and disappears. Even were the birds to return they would not be able to stop this unraveling of the community and transition to a new state. The forest canopy continues to open until all trees have disappeared, leaving only shrubs, a process we call a *positive feedback*—it simply gets worse and worse. We have photographic records of the decline of this forest in many places on the Mara River since 1966—a stable forest state moving to a stable nonforest state. At this point we are not clear on how the reverse process works—the development of forest—because it is very different from that in the lowland forest of the Grumeti River (mentioned in chapter 7).

The stable forest state is also an example of what we call a top-down cascade. In essence, birds depress beetles indirectly, which promotes seedling regeneration and a stable, closed-canopy forest. Remove the top predators (in this case birds), and the forest collapses. But the example is even more complicated because the birds acting as tree-seed dispersers produce a mutualistic interaction promoting stability of the forest ecosystem (see chapter 7 for a discussion of mutualism), and the change in state is as much due to the breakdown of this association as to predation.

———

These few examples of multiple states in the Serengeti ecosystem bring out two important points. First, we see that a minor disturbance can be tolerated without radically altering the community of species. But

if that disturbance becomes too severe, the community changes, usually by a positive-feedback unraveling, to another set of species. In our
first example, excessive burning, starting in the 1920s, eventually led to
a woodland with mature trees but with no seedling replacements—
this was the beautiful parkland that conservationists and tourists met
and cherished when these activities began in the early 1960s. It was
considered to be the natural state of African savanna,[13] but in reality
it was entirely unstable, and the mature trees eventually died of old
age within a short period of time, as aerial photos from the 1950s and
'60s showed, leaving only a grassland. In the second example, we have
less concrete evidence, but the modeling and circumstantial data suggest that the massive die-off of wildebeest due to rinderpest in the
1890s and over the next 60 years allowed predators to kill enough
calves to keep the wildebeest population at a very low level (top-down
regulation). Then, when rinderpest was removed, wildebeest numbers
jumped (because of the increased survival of calves) to the new level
regulated by food (bottom-up regulation). In the third example, the
same excessive burning disturbed the forest canopy, destroying
the stabilizing processes of bird frugivory (top-down) and unraveling the
community.

The second important point is that predators are one of the fundamental causes of multiple states. The theory of predation (see note 3 in
chapter 5) predicts that predators can exist at two levels of prey-
population size: a high level of prey when predators are not limiting
prey numbers and a low level when predators do limit prey numbers.
Here we see that elephants, as predators of trees, can exist with many
trees and with very few trees. With wildebeest, predators feed on the
large migratory population but cannot limit the numbers, but very low
wildebeest populations may indicate that predators have been able to
regulate them.

Multiple states may not be common, but we see examples in grasslands, forests, lakes, rivers, and coral reefs. In chapter 12 we will see examples where one of the states is undesirable.[14]

———

In essence, if a disturbance is too severe or too persistent, the ecosystem may switch rapidly to a new configuration of species, a new state. Communities can exist in different states when different combinations of species abundances occur under the *same* environmental conditions. Usually changes in state result from top-down effects of predators, but they can also result from changes in competitive interactions among species, and from environmental or human disturbances. In the context of regulation, we can say that equilibrium can jump suddenly between states. This showed me how complicated regulation can get: there was a switch in the direction of regulation.

11

The Fundamental Principle of Regulation, and Future Directions

When I set out to study the Serengeti in the 1960s, I asked what made Serengeti so outstanding compared to other ecosystems in the world: What were its special features and what were the conditions that caused it to remain this way at least over near time—the past 1,000 years or so? Over five decades the answers have unfolded in a series of steps, each one built upon the previous, showing progressive complexity, which in the end exposed the principles underlying how ecosystems work.

The first question concerned the issue of regulation: Are populations stabilized, or regulated, by processes that normally prevent overabundance or extinction? The fortuitous events surrounding the disappearance of the great disturbance known as rinderpest allowed me to make the critical test for regulation and demonstrate that it did indeed occur in both populations that I studied, buffalo and wildebeest, involving changes in the rates of death or birth. *Regulation of populations* results in an equilibrium population size that may or may not be seen in reality (chapter 3). Regulation is a fundamental principle that has emerged from this work and appears to apply to all populations. It is the reason why the ecosystem persists in nature, at least in near time. This **fundamental principle** is the basis of all others that are modifications of it, which I was to uncover later. These I will call **subprinciples**, of which there are seven.

Regulation is caused by changes in birth and death rates; in buffalo and wildebeest, it was death rates that changed the most as numbers increased. Driving the increase in death rates was a progressive increase in starvation as food supply (green grass in these cases) ran out: there was less and less food for each individual from a constant pool of food available in the dry season, the season of short supply. Eventually the percentage of the population starving matched the proportion entering through births, stabilizing the population. This feature was clearly seen with the wildebeest numbers that leveled out at about 1.3 million after 1977, remaining at that level for the next 40 years. Thus the resources that normally regulate population are food supply or space for territories. Too many animals results in less food per individual (or space within which an animal can get food), and eventually it dies. Too few animals mean more food, greater survival and reproduction, and ultimately population increases. Thus, regulation comes from below, from the plants, and this **first subprinciple** is called *bottom-up regulation*. It applies often with the very large mammal herbivores, not just in Serengeti but around the world (chapter 4).

Bottom-up regulation, however, applied to only a few of Serengeti's herbivore species. What was regulating the other, smaller species? The evidence did not indicate that food was in short supply; animals were not starving. Instead, the evidence showed that predators were regulating their numbers. These small herbivores were less concerned with finding food than with avoiding predators, the most frequent cause of their deaths. In a fortuitous experiment, numbers of prey species were found to increase when predators were removed, and to decrease again when predator numbers rebounded. Predators can keep populations at a lower equilibrium than would occur through resource limitation, eating more if numbers increase, less if they decrease. Regulation, therefore, comes from the trophic level above, which is the **second subprinciple,** *top-down regulation.* It is especially important because it seems to apply to the majority of animal species in ecosystems around the world (chapter 5).

All mobile animal species move with the seasons, but some move much greater distances, and these are called migrants. Why do some species move such great distances, and why do most of these have large populations? In Serengeti populations of migrant wildebeest, zebra, and Thomson's gazelle number in the hundreds of thousands, but nonmigrant populations of the same species in the same area number in the few thousands only. And this phenomenon applies to many other species of mammals and birds in Serengeti and elsewhere. It turns out that migrant species differ from resident species in being able to access temporary but highly nutritious food because of their ability to move. They arrive when food is available and leave when it disappears. Because migrant species are bottom-up regulated (i.e., regulated by food abundance), this extra temporary food allows great numbers. Migration also results in less predation because predators cannot migrate with their prey, which also contributes to greater numbers. Conclusion: *migration raises the equilibrium level* through bottom-up regulation, our **third subprinciple** (chapter 6).

There is a huge diversity of species in Serengeti, whether it is mammals, birds, or insects. Are all these species being regulated independently of each other, a sort of random collection of populations going up and down? Or is there some connection among species that modifies the type of regulation each species experiences? The evidence shows that the diversity of sizes of various herbivore prey species, in tandem with the diversity of sizes of predator species, determines whether a species is top-down or bottom-up regulated. These two processes together cause less fluctuation in numbers of either prey or predator than if there had been only one or two species each (as we see in fluctuating populations of northern Europe and North America). Basically, this diversity of size supports a general hypothesis that the more species there are interacting together (feeding on, competing with, or supporting other species), the stronger the overall regulation of the whole community is.[1] This hypothesis proposes that the *higher the biodiversity, the stronger the regulation* and the less disturbance around the equilibrium— our **fourth subprinciple** (chapter 7). Of course, not all species are equally important in creating stability; the most important ones, the *keystones*, play a vital role in creating the biodiversity and hence the stability, a top-down effect. But other connections—mutualism, commensalism, and facilitation, which are predominant interactions in Serengeti (chapter 7)—also hold communities together.

The Serengeti environment fluctuates from year to year largely through changes in the amount of rainfall. These variations affect all other components of the ecosystem: changes in growth of plants and hence food for mammal, bird, and insect herbivores, as well as birth and death rates of animals and hence their population sizes, and the latter translate into changes in food and numbers of predators. Many of these fluctuations in different parts of the ecosystem are connected because they are all fundamentally caused by changes in weather cells from El Niño in the Pacific Ocean. Connections among components provide stability in themselves, but the underlying regulatory mechanisms provide the main stabilizing process. So *environment disturbs the populations*, and regulation then corrects these disturbances, the **fifth subprinciple**. Such disturbances are the reason that equilibrium is not usually seen in the form of a stationary population.

Despite the evidence showing how the Serengeti ecosystem is regulated through several different complex mechanisms allowing persistence of the system (subprinciples 1–5 above), it became clear over the decades that Serengeti was changing in most aspects of environment and biota—climate, plants, and animals. What caused these changes, and how do they relate to regulation? Over long periods of Earth's history we see that its environment is constantly changing. Compounding this change over the past 200 years since the Industrial Revolution began— but vastly more pronounced in the past 50 years—are the effects of human-caused climate change. Together these environmental changes mean that within any ecosystem the climate, the weather disturbances, and the level of resources also change continuously. *The equilibrium level of the system is always changing*, so that regulation is always tracking this changing level. This is the **sixth subprinciple** (chapter 9).

Research in Serengeti has now shown that some components have changed but never returned to their original form. Dense forests and thickets that once covered almost the whole of the northern Serengeti and Mara Reserve 100 or more years ago have progressively disappeared and been replaced by grassland. Savanna in the Mara Reserve some

70 years ago has also become grassland, never to return. What has produced this alternative state? Serengeti has evidently been changing, but if this or indeed any other system changes too much, or if the disturbances become too great, or even too small, the system can flip into a different state, with a different community of species. And this different state cannot always be changed back to the original. Savanna occurs within a range of rainfall per year; if the environment becomes drier, savanna changes to grassland, if wetter, it changes to dense woodland. If disturbances are too severe—say, there are too many fires—then savanna cannot regenerate; it changes to grassland and cannot change back again because elephants keep it as grassland. If not enough burning occurs, savanna changes to dense thicket, which stays that way because fire now cannot penetrate. Consequently, *if an ecosystem is disturbed too much, it can flip into an irreversible state*—our **seventh subprinciple** (chapter 10). The equilibrium can jump suddenly to different levels.

Where Next? The Trophic Cascade

I have focused thus far on the ways in which processes drive the dynamics of the food chain, either from the bottom up or from the top down, or in various combinations. I have labeled these processes subprinciples. But these are by no means all of the food-chain interactions in the Serengeti ecosystem—there is still much to be elucidated. I have concentrated on interactions among predators, prey, and plants. Little has been said so far on the still lower levels of minerals, soils, and the interaction of both with plants. We know that climate change is altering the availability of water to the system, and water flow in rivers is changing (see chapter 12). How changes in hydrology affect nutrients and plant production clearly needs examination. The work of Han Olff of the University of Groningen, as well as that of Grant Hopcraft and Kristine Metzger,[2] has shown that underlying parent geology (the underlying geological material) and the resulting minerals and soils vary across the Serengeti. Apart from the Serengeti plains, with their high volcanic-mineral content (described in chapters 1 and 6), there are areas of high nutrients in the savanna—such as the corridor (the western spur of the ecosystem between the Grumeti and Mbalageti rivers) and northeast around Lobo—and areas of low nutrients, particularly in the broad-leaved woodlands of the northwest, the Maswa Game Reserve, and on the hills, derived from underlying granites. Plants respond to these underlying rock differences: in the granitic northwest, the grasses show a different species composition from that in the volcanic areas. The grazers in turn respond to these differences in the type and quality of the grasses; resident densities of ungulates are much higher in the higher-nutrient western corridor and northeast than in the low-nutrient areas. This is obviously a bottom-up process.

But there is also a top-down pressure. The extensive work of Louis Verchot, Naomi Ward, and others[3] has shown that different intensities of grazing result in different soil biota and soil-nutrient content. Persistent heavy grazing, as occurs with cattle herding, results in a fungal-driven soil biota, lower densities of nematode worms (essential in decomposition), and lower nitrogen and carbon in the soil. In contrast, seasonal grazing, resulting from the periodic appearance of wildlife migrants, produces a bacteria-dominated soil biota, higher nematode numbers, and high soil nitrogen and carbon. This top-down effect of heavy grazing then feeds back to produce a plant community dominated by low-nutrient flowering plants, rather than the high-nutrient grass communities in seasonal grazing regimes.

Michiel Veldhuis and colleagues[4] have shown that persistent grazing is now becoming more prevalent within the Serengeti ecosystem, not just from the encroachment of cattle, but also because wildebeest are forced to graze more intensively in their own areas, as shown by Tom Morrison and colleagues (which we return to in chapter 12).[5] Future work should look at the possibility of a shift in trophic dynamics from a nutrient-rich to a nutrient-poor system, with all the consequences resulting from such a bottom-up flow, not the least of which would be a decline in wildebeest numbers.

Mike Anderson and Tom Morrison have examined the factors that allow tree-seedling germination. They found a complex interaction of factors in the tall grasslands where trees grow—on the savanna, rather than the plains. These tall grasses effectively prevent germination of seedlings through competition. If seedlings are to succeed, a series of special circumstances must prevail: the abrasion of seed coats from being eaten by browsers (impala like eating acacia pods, for example), which enables germination; heavy rainfall to moisten the soil for early growth; and removal of grass competition by heavy grazing, followed by dry conditions during the dry season to keep out grasses and further reduce competition. Essentially, this involves both a bottom-up process (soil moisture) and a top-down process in which herbivores play a double role—a mutualistic one when they digest the seed pods and an indirect predation one when they graze down competitors. Nevertheless, even when wildebeest grazing reduced grass competition as the population increased, these researchers found that seedling germination remained difficult, slow, and relatively infrequent.[6] Yet our evidence from the photographs showed a rapid mass regeneration of small trees, which started over a short period of a year or two (figure 9.2). How do the germination experiments mesh with the observations from photos? One possibility is that the regeneration that took place in the 1970s and '80s was not from seedling germination—it was from resprouting of well-established rootstocks that had been burned back to ground level year after year from the extensive hot fires over a 40-year period. Some of these stocks were as much as five centimeters across, others more like one to two centimeters. They had germinated many years previously and were just surviving belowground under the heavy burning regime of the 1950s and '60s. But this does raise the question: When did these rootstocks become established, and what allowed them to germinate in the past and persist for so long? Tom Morrison and Mike Anderson have also found that in the 2010s there was a distinct difference in the tree–shrub species composition of overstory and understory trees.[7] The dominant overstory trees are the umbrella acacia (*Vachellia tortilis*) and stink-bark acacia (*V. robusta*), whereas in the understory there is a predominance of small species such as caterpillar bush (*Ormocarpum trichocarpum*), a two- to three-meter, corky bush, and sicklebush (*Dichrostachys cinerea*), a three- to four-meter, thorny small tree, while regenerating overstory species are in a minority. Dennis Herlocker, our plant ecologist in the 1970s, found that in 1971–73, prior to the increase in overstory trees (1977–2005), the dominant small trees were gum acacia (*Senegalia senegal*), *V. hockii*, and *Commiphora*.[8] The presence of this understory is associated with high-density soils, a bottom-up process, rather than with fire and grazing. In addition, Morrison, and colleagues[9] have found that with the increase of elephant numbers in the 2000–2020 period, their browsing was the major cause of mortality of sapling trees (taller than two meters); elephants are selective in their diet and so determine which species make up the adult-tree community, a pure top-down effect. These changes highlight the need to understand the dynamics of tree composition and succession in future work. It also highlights the need to study the effects of fire. Stephanie Eby has presented a robust analysis so far,[10] but there is clearly more to the story.

I described earlier how ungulates divide up habitats and food to reduce competition among the species. These studies were relatively unrefined because of the limitations of methods 50 years ago. Anderson, Olff, Eby, Hopcraft, and others have now used more refined methods to show how landscape and habitat, disturbance from fire and grazing, food nutrient content, and risk of being caught by natural predators (and humans) all interact to determine in which habitat ungulate species choose to live.[11] Cameras set on posts to photograph animals that pass by are the developing method for future studies of animal movements, habitat preferences, and species associations. Craig Packer and Mike Anderson set out a grid of over 200 camera traps in a 1,000-square-kilometer portion of the Serengeti plains to study which ungulates live in which habitats and which predators occur with those habitats and ungulates.[12] Lydia Beudrot used these photos to test our earlier ideas on whether competition for food or avoidance of predation explained ungulate associations, and she found that when food is plentiful, risk of being eaten determined the associations. Ali Swanson used these photos to complement the findings of Grant Hopcraft, who is monitoring the movements of the migratory wildebeest and zebra with respect to where food, natural predators, and humans occur. He finds that wildebeest move according to where food is found and tend to ignore threats from predators. In contrast, zebra are more sensitive to where predators and humans occur. These are the new ideas on dangers that are being explored; they should help with conservation priorities, which I address below.

Craig Packer has shown that the top-down regulation of smaller carnivores depends on the species. Lions and hyenas, although they kill some infant cheetahs, have not had an impact on that species' numbers. In contrast, these top predators kill young wild dogs and steal the food; predation is partly why wild dogs went extinct within the Serengeti Park.[13] The top-down impact on other small carnivores is unknown, making this a major area for future investigation.

Another new development is the use of DNA analysis to detect which plant species are eaten by ungulates by testing samples of their feces. Rob Pringle and Mike Anderson are examining how the diets of migrant species change as they move across the landscape. As they expand this DNA approach to more species, we will get a greatly refined, more detailed picture of what herbivores eat and how their food niches are separated—how bottom-up processes of competition divide up the food niches.

The role of insects, despite their ubiquity and vast variety, in both bottom-up and top-down processes is almost entirely unknown. The exceptional work of Sanne de Visser and Bernd Freymann of Groningen University, using the insects' chemical signatures to show what plants they feed on, has provided valuable information on the invertebrate food chain.[14] This approach is clearly the way of the future, along with DNA analysis. Early data from 1975 suggest that grasshoppers can eat as much as the wildebeest in the wet season, but in the dry season they die and have little impact.[15] Bernd Freymann has shown how termites affect the nutrient content of soils in the vicinity of mounds,[16] and Ally Nkwabi has demonstrated a possible top-down effect of birds on grass-layer invertebrate populations.[17] But these are just the start, and the study of invertebrate-community dynamics is a field ripe for new work.

There is one area of multiple-state studies that requires exploration, not just for understanding ecosystem dynamics, but also for conservation. Predators have the ability to reduce prey populations, as described in chapters 5 and 10. However, there are some situations in which the

prey population is stuck at that lower level, whether due to predator behavior or habitat and climate change, and the prey may even go extinct—a
process called the *Allee effect*. We know *how* this occurs, but we do not
necessarily know *why* it should occur—and the answer to that question
is vital for conservation.

One of the clearest illustrations of the Allee effect is the plight of the
right whale in the North Atlantic Ocean. Its population is so low that it
cannot increase fast enough to counteract random killings from ship
collisions; it is stuck at these very low numbers and is in danger of extinction. A thousand years ago the right whale was found throughout
the North Atlantic, but excessive hunting between the thirteenth and
seventeenth centuries reduced it to such low levels that by 1700 whaling
had become unprofitable.[18] Despite being protected for many decades
in modern times, it has never been able to increase again. The question
is why. Similarly, in Serengeti, the rhinoceros population, which num-
bered some 500 animals in the 1970s, was virtually exterminated by illegal hunting for their horns over a two-year period, 1977–78. A few animals have wandered back into the park and have been carefully guarded
over the past 30 years, but their population seems to be stuck at about
30 animals. Reproduction is far lower than one would expect of rhino
when food is abundant, which it is. Surprisingly, these few rhino are
genetically diverse, so inbreeding does not appear to be a problem.

Another subject where research is needed is the role of disease as part
of the trophic cascade; this area has hardly been touched. Ricco Holdo
and others have confirmed the top-down cascade resulting from the
imposition of the exotic virus rinderpest, which affects all other trophic
levels, with consequences extensively documented in this book.[19] This
is one of the first demonstrations of disease as a regulating agent. The
role of the huge number of endemic diseases that affect wild ungulates
may well be subtle; when mammals begin to starve, their immune systems are compromised, and one of the many diseases and parasites normally residing in the animal can take over and become pathological.
Which of the parasites takes over may well be a lottery, so that a range
of species are involved with the regulation of the host.[20] It is highly
likely that the effect of disease is to speed up starvation, making bottom-

up regulation sensitive and fast acting. I can point to a test of this idea with the water buffalo of Australia. These animals, closely related to African buffalo, were released into northern Australia from Asia in the mid-1800s without their normal burden of parasites. The population increased to high numbers by 1972 on the floodplains of the Northern Territory, and they began to overgraze the land. Without their usual parasites to help control their population, they were able to continue to feed until there was no food remaining, just bare ground. Then they starved until they became walking skeletons and died *en masse*.[21] Parasites may well be an essential component of bottom-up regulation, and their absence could lead to greater fluctuations in population numbers.

One last aspect needs intensive research. Humans have disrupted, even destroyed, native ecosystems by their unregulated exploitation, especially via agriculture but also by other means, such as forestry practiced over many centuries without regard to its effect on ecosystem processes and loss of biodiversity. These impacts need to be examined to determine whether they could lead to failure of systems vital to humanity, and also to determine whether the impacts could affect the long-term persistence of native systems. These studies need a baseline, which we propose could be provided by protected areas such as Serengeti (where human impacts are much reduced compared to those in, for example, the agricultural areas immediately adjacent to them); these baselines would act as standard scientific controls. This aspect I will deal with in chapter 14.

Apart from these outstanding components of the food chain that still need to be explored, there are a number of other discrete topics that require future study. First, there is the question of why there is such high local endemism in birds—that is, local populations separate from the rest of the distribution, subspecies, or even full species. This is a feature of Serengeti that even the field guides have noted. Normally such locally distinct populations evolve because an area has been cut off in the past. Whether this has occurred, and if so how, remain open questions.

Second, despite the considerable research in Serengeti, several species have been neglected. Perhaps the pangolin is the most urgent; given that this is the most endangered group of mammals in the world, we know nothing, and we need to know a lot. Surprisingly, hippos are almost completely unstudied, despite their large numbers and considerable grazing impact on grasslands. They turn long-grass tussock grasslands into short-grass swards, their effect stretching some 20 kilometers from the rivers where they rest during the day. Their nocturnal habit is one reason little is known, as researchers are not encouraged to roam the country at night. For the same reason, we know little about other nocturnal species, especially small carnivores such as the African wildcat, white-tailed mongoose, aardwolf, and aardvark, as well as nocturnal ungulates such as dikdik and reedbuck. Daytime observations show that reedbuck live along rivers, although recent nighttime drives have shown a far larger population not previously known to exist throughout the long-grass plains away from rivers. Many of these species are now coming into view through the camera traps set up by Ali Swanson, Craig Packer, and Mike Anderson—as noted, this is the way of the future.[22] There are also some anomalous plant distributions that have yet to be explained: the stink-bark acacia (*V. robusta*) is the dominant savanna species over most of Serengeti, but it is replaced by the wadi acacia (*V. gerrardii*) in a narrow band across the northern Serengeti near the Grumeti River, and this species then extends across the whole Maasai Mara Reserve. What causes this switch needs further investigation. At the same time there are two morphs of stink-bark acacia, a dominant one that has a solid, spherical canopy, and a flat-topped one that resembles forms in southern Africa. The two morphs can live side by side, and botanical specialists proclaim them both the same species. Why do we have two morphs in such close proximity?

We have come a long way over the past 50 years in understanding how an ecosystem works through variations of the principle of regulation. Nevertheless, there are still lots of new questions to be answered, and the answers will likely cause us to rethink the causes of the changes we have described. These are for future generations of Tanzanians to explore.

12

Threats to the Serengeti

The seven variations of regulation, what we have called the subprinciples, have important consequences for conservation in general and
Serengeti in particular. To start with, if the **fundamental principle of
regulation**, which stabilizes a population, is distorted, then regulation
breaks down. When the great drought of 1993 caused massive starvation
and death in wildebeest and buffalo, there was an increase in the survival of calves in following years, which allowed both populations to
recover. This is regulation as it should occur. However, we see a quite
different result with elephants. From about 1950, numbers increased
until the mid-1970s, reaching 3,000 in the ecosystem. After the border
closure in 1977, ivory poaching increased and persisted for a decade,
causing a collapse of the elephant population. The most severe poaching
took place in the 1980s in Serengeti, when some 80 percent of the population was lost (see figure 4.6). During this period, when numbers were
very low, we should have expected an increase in the birth rate to compensate for the increased deaths, as we saw in other species. In fact, we
saw the opposite; during the poaching, which was clearly an extremely
stressful period for the elephants, females nearly stopped producing
babies for 10 years. There is some evidence that this was a stress effect.
Scientists from Norway studying elephants in the 2000s were able to
obtain in Serengeti blood samples of elephants that had the habit of
creeping outside the park to raid agricultural crops, especially corn and
millet. The scientists found that stress hormones increased markedly in
animals outside the park boundary, and even inside the park as they

approached the boundary—elephants know that humans are stressful.[1] The important conservation consequence is that at very low population numbers, under extreme stress (as these elephants were in the 1980s), the population not only could not recover but was in serious danger of collapsing to extinction because of the drop in reproduction. This may be one mechanism producing the Allee effect discussed in chapter 11.

Interfering with Bottom-Up Processes (Subprinciple 1)

The overall abundance of resources determines the abundance of the various species that use them. As noted in chapter 4, these resources— water, food, space—set the baseline or equilibrium position of a species (**subprinciple 1**). If these resources for the population are reduced, there is less to go around and numbers will decrease, particularly when species are regulated by bottom-up processes. Several such threats related to resources are looming.

The most important of these is the threat by authorities to dam the Mara River in the Mau Highlands of Kenya. The Mara River is the most important water source for the Serengeti migration in the dry season. The animals move north to the Mara because it is the only flowing river of any size in the dry season and provides vital water for the millions of animals. But the Mara has its origins in the Mau forests of the highlands of Kenya, far from the Serengeti ecosystem. A dam would certainly reduce the flow of water in the river downstream during the dry season. And water catchment is being further reduced because the Mau forests are being cut down at an accelerating pace, which alters the seasonal flow of water; without forest trees, more rain runs away in the wet season, causing floods. Less rain then percolates through the ground into the rivers in the dry season, and less enters the river. A dam would stop even this decreased flow and impound the water. The situation is further compounded by water off-take downstream. The Mara flows through agricultural land where unregulated irrigation is tapping off the water. As a result, the Mara has in the late 2000s stopped flowing for the first time in a century. These impacts take place in Kenya, out of reach of controls from Tanzania, meaning that unless agreements are drawn up to regulate the irrigation off-take, particularly in the dry season, the wildebeest could find there is no water for them when they arrive in Serengeti in the next few years.[2] Without the Mara River, the wildebeest migration would have to find an alternative source, and the only possibility is Lake Victoria at Speke Gulf. That would change their migration route, likely resulting in declining wildebeest numbers, in turn affecting numerous other species. For Kenya, this route change would radically reduce the tourist value of the Mara Reserve because there would be few animals going that way.

We already have clear evidence of the change in flow of the Mara River as a result of increased wet-season floods over the past 50 years. The width of the river has almost doubled since 1967,[3] and consequently forest along the main riverbed has been washed away. This bank erosion has been caused by more voluminous floods in the wet season, most likely due to the extensive deforestation of the Kenya highlands since 1975, because rainfall has not been higher in that period. Consequently, the extent of evergreen groundwater forests along the Mara has been reduced since the 1950s.[4] This raises a conservation concern for the loss of even more forests in the future, because they support many animal species confined to this habitat. The problem is exacerbated by the fact that the same forests have been completely eradicated outside the park. Examples of affected animals include some birds, such as a subspecies of Schalow's turaco, and the subspecies of black-and-white colobus monkey, both confined to the Serengeti region.

A corollary to the reduction in water flow on the Mara River concerns Lake Victoria itself. Lake Victoria is a vast but shallow body of nearly 70,000 square kilometers immediately to the west of Serengeti. It is so large that it creates its own weather system, producing rainstorms in the dry season in the west and northwest of the ecosystem. These storms are really an extension of the Congo rainforest weather systems, and they are the same storms that provide the food for the wildebeest migration. The lake has only one major water inlet, the Kagera River, on the west side, as well as some smaller rivers on the east side, the main one being the Mara. If the Mara stops flowing, the lake—only a few meters deep around its shores—could recede as water levels drop. On Speke Gulf, the shores could move several kilometers back from their current locations. Millions of people dependent on the lake for fishing, irrigation, and drinking water would find themselves stranded. In addition, the weather systems of the lake could be changed, with dry-season rain declining as the lake receded. And this would affect the migration pattern of the wildebeest as well.

We should take to heart the lessons from the drying up of the Aral Sea, in Kazakhstan, the consequence of uncontrolled off-take of water for irrigation. In the 1960s, water was diverted from the two rivers feeding the Aral for cotton and rice agriculture. By the 1990s, the lake was dry, freshwater fisheries had collapsed, and many people had lost their livelihoods. Then extreme weather, brought on by the dried-out Aral seabed, compounded the misery of the people. To make matters even worse, irrigation resulted in salinization of the soil so that the crops failed. Remedial action has now been taken; part of the lake has filled and salt-tolerant fishes introduced, effectively making it an artificial system.[5]

Another impact of the limitation of resources through bottom-up processes (**subprinciple 1**) concerns the imposition of boundaries around the ecosystem. A century or so ago, animals in Serengeti were able to move as they wished to follow the scattered rainstorms in the dry season; travelers reported wildlife well to the west of the current boundary, especially the migrants (see chapter 9). Michiel Veldhuis, Han Olff, and colleagues have documented the increasing pressure from illegal grazing by Maasai pastoralists on the eastern boundaries within the park, which has now changed savanna to dense thicket. As a result, the migrant species avoid these areas and are confined to a smaller area within the center of the park.[6] Essentially, the food available for the wild animals has decreased, and I predict that wildebeest numbers will drop. The equilibrium level will have declined.

This raises the general problem for conservation that because protected areas are mere samples of the original native habitats, they cannot contain some of the species they should protect. Some species have far greater ranges than can be contained within a park. For example, a male elephant was tracked by Alfred Kikoti from Mt. Kilimanjaro west across the Rift Valley and into the Serengeti, a distance of several hundred kilometers.[7] Emmanuel Masenga and Ernest Eblate, of the Tanzania Wildlife Research Institute, have translocated several wild dog packs from outside the eastern Serengeti and released them in the corridor. Some of these packs have remained in Serengeti, but one moved north almost as far as Nairobi, then southeast to Tsavo National Park, then south again to eastern Tanzania (500 kilometers from Serengeti), before returning to somewhere near the Mara. Clearly this species is wide ranging and cannot be kept in one area.[8] Boundaries, therefore, become a limitation on the space that wildlife would use, a limitation that will affect population sizes in the future. Conservation must recognize the need for these movements and find ways of facilitating them—a daunting task, but one that should be examined.

On the same theme, protected areas are simply islands in a sea of agriculture or other highly modified landscapes unsuitable for native species. This means that plants and animals are confined to population sizes that are constrained within the park, and many of those populations are simply not large enough to allow long-term survival; they will go extinct eventually. (This is known as the *extinction debt*—species still living but doomed to extinction). Many bird species, for example, require large populations to survive. Other birds live in specialized habitats that have disappeared outside the park, and what remains inside is too small; riverine-forest birds fall into this category. Conservationists recognize the need for corridors of habitat to join protected areas as a network. How to do this in human-dominated landscapes is the big problem.

One especially important area for future work is how to engage with local peoples so that they can understand both the economic and environment benefits of such protected areas. Human–wildlife conflicts are increasing as humans are abutting the protected-area boundaries. One approach has been to construct fences to keep wild animals out of agriculture.[9] Unfortunately, this has considerable disruptive effects on the trophic cascade, resulting in distortions of species populations: too many predators, destabilization of prey populations, overgrazing of plant communities, and soil disruption, among other consequences, as Sarah Durant, a leading carnivore ecologist in Serengeti and conservationist at the Zoological Society of London, and a multitude of others have shown. This simplistic solution results in parks becoming mere zoos, which defeats the purpose of protected areas as ecological baselines, an aspect discussed in chapter 14.[10] An alternative has been suggested by local Tanzanian administrators and politicians. Because tourist lodges and hotels inside the park are now becoming excessive and resulting in major disturbances, these senior officials have suggested moving hotels outside the park, to a buffer zone, where local peoples can benefit from employment in the lodges and from sale of their produce. At the same time, agriculture in such buffer zones would be removed, but livestock use encouraged. This simultaneously reduces human–wildlife conflict through a gradient of human and wildlife use, and removes the necessity of fences. Human use predominates at the outside end of the gradient, wildlife at the inside. Predators can then have their normal prey instead of domestic prey. Programs to trial this initiative are urgently needed.

In 1988, ivory poaching ceased when elephants were given endangered-species status and trade in ivory was banned. Subsequently, the elephant population increased rapidly to its numbers of the 1970s, and animals began moving outside the park boundaries, as they used to do. But by then, in the 1990s, the human population had increased hugely (some 400 percent since the 1960s) and had converted the areas next to the park into intensive agriculture, with crops that elephants love to eat. Humans are now encroaching on the ecosystem in Maswa Game Reserve with agriculture, and on the eastern boundary through excessive cattle grazing. The result is conflict between elephants and humans over territory. Space resources that wildlife used to possess have been taken away.

Tourism, as noted above, is another aspect that affects the resources of the ecosystem. The two major impacts are the building of hotels and lodges, and the use of vehicles to observe animals. There are some benefits from tourism: it brings in revenue and provides economic justification for the existence of the park. The continuous exposure to vehicles results in animals becoming habituated and so less disturbed, less stressed. We can see this around Seronera with its tame buffalo herds, which allow vehicles to watch these beautiful animals close up. At one time, herds turned and ran at half a kilometer's distance and all you saw was a cloud of dust. Similarly, elephant groups now continue their normal social behavior in close proximity to vehicles. An important positive outcome that we hope will materialize is a drop in poaching activity in the far northwest of the park around Kogatende and Lamai due to the presence of lodges and vehicles. Poachers find it difficult to operate when tourists are watching (although they still operate at night).

But tourism also has negative consequences for conservation. One is the soil disturbance and erosion caused by the tracks used by vehicles. Obviously, we should expect some "sacrifice areas" (a term used in agriculture for disturbed ground near gates, feeding barns, etc.), but with too many vehicles the soil disturbance becomes severe and lasts for decades. The north-south park boundary across the eastern plains from near Barafu to Lake Lagarja has been used by vehicles since the 1960s, and we can count at least 50 parallel old tracks—a considerable disturbance. Less but still noticeable soil disturbances are seen along the heavily used tourist roads around the main rivers such as the Seronera River. In these high-use areas, the days of tracks should be gone, as is the case in Kruger National Park, South Africa, where only hardened roads are now used. In addition, the number of vehicles is now so great that traffic jams occur around lion prides, and vehicles crowding around wildebeest crossing points on the Mara River actually prevent wildebeest from crossing.

Lodges and hotels take over habitat, often special ones such as kopjes, thus preventing use by wildlife. A few buildings are not a concern. The conservation problem is that there is no stopping rule. Thirty years ago, it was thought that three hotels were all that should be allowed in Serengeti. Then the number was increased a decade ago, and now lodges are being constructed at an accelerating rate, with all the attendant disturbances of tracks and vehicles. When there is too much disturbance to animals, they cannot live naturally. Grazing is inhibited, predators cannot hunt, wildebeest cannot move naturally. We see this in the extreme form in the Mara Reserve of Kenya, where there is no control on tracks, vehicles, or visitor numbers. Joseph Ogutu and his team have shown that resident ungulate populations have declined significantly since the 1980s, partly due to excessive tourism and partly to illegal hunting and livestock grazing. If the migration changes route, as we discussed above, the Mara Reserve in Kenya will cease to be a fulfilling wildlife experience for tourists.[11]

We are now seeing the negative impact of lodges on wildlife in Serengeti National Park as well as the Mara Reserve. Tom Morrison and Grant Hopcraft, using radio collars on wildebeest and zebra trackable by satellite, compared areas of the Serengeti ecosystem used in the early 2000s with those used 20 years later.[12] They and colleagues found that both species had changed their grazing areas from ones they preferred for their superior food to more peripheral zones with poorer food so as to avoid tourist lodges; unfortunately, the lodges had been placed in the best areas. Since the number of lodges and hotels is continuing to increase, this negative effect on migrant ungulate behavior will get even worse—and with it all the trophic consequences of declining migrant numbers on the entire ecosystem. Obviously, there has to be some moderation of tourist interference, and one suggestion, outlined above, is to remove lodges to the edge of the park, where their disturbance is minimal. Further studies and trial solutions are important new areas of research.

Interfering with Top-down Processes (Subprinciple 2)

There are top-down as well as bottom-up consequences for conservation. In a natural system, top predators regulate smaller resident ungulate species, and herbivores such as elephants limit plant species (**subprinciple 2**). If another top predator is imposed on a system, the trophic cascade is distorted and normal top-down regulation may not occur. This happens when humans impose heavy hunting pressure (we call it poaching if it is illegal) on resident ungulates, elephants, rhinos, and lions. In the 1970s and '80s, such heavy poaching took place that all rhinos were exterminated, 80 percent of the elephants disappeared, and 85 percent of the buffalo were killed. In northern Serengeti, buffalo and elephant were even exterminated. Most of the illegal hunting now focuses on the migrants, but many of the residents are caught up by mistake (called *bycatch*), and their populations have been driven down almost to extinction, especially in the western corridor. Topi populations declined markedly due to poaching; the proof of this came when major antipoaching operations were instituted in the privately operated Grumeti Game Reserve in the 2000s, and topi numbers returned to where they were in the 1960s. The conservation concern is not that human hunting takes place—some species can tolerate some hunting off-take—but that it is not controlled. It is determined by the number of people on the boundaries of the park, and this number is increasing rapidly. If this continues, wildlife populations will collapse.[13]

Disease is a top-down process and a major concern for conservation and surrounding human societies. Sarah Cleaveland and a large group, including Ernest Eblate, Tiziana Lembo, Katie Hampson, and Meggan Craft, based at the University of Glasgow, have explored the impact of various wildlife and domestic-animal diseases. This group has found that canine rabies, lethal to humans, is maintained in the domestic dog population, which has been increasing rapidly around the borders of Serengeti and transmitting the disease to both humans and wildlife. Sarah has found from a widespread vaccination campaign of dogs that rabies can be controlled. The objective now is to eliminate the disease by 2030.[14]

Sarah Cleaveland's disease team and Craig Packer's carnivore team have studied another important wildlife disease, canine distemper virus (CDV), which killed many lions in 1994. They find that once again this virus is transmitted to carnivores, including both lions and hyenas, by domestic dogs. The Serengeti research suggests that, in contrast to rabies, dog vaccination alone is not likely to be effective in protecting wild carnivores such as lions and wild dogs against CDV outbreaks; there is still sufficient residual transmission to kill wild carnivores. Instead, direct vaccination of endangered wild species, even at low coverage, can better protect against CDV extinction threats.[15]

Both rabies and CDV involve the transmission of disease from domestic animals to wildlife. The next two concern the reverse process. Foot-and-mouth disease (FMD) is an economically important disease of cattle. In southern Africa, the source is thought to be African buffalo, and fences have been constructed to separate the two species—with all the negative consequences of fences discussed above. Serengeti buffalo, it turns out, are only minor reservoirs for FMD; thus new programs are being designed to vaccinate cattle instead.[16] Malignant catarrhal fever (MCF) is a fatal disease of cattle transmitted from wildebeest via the fetal membranes left behind from the birth of calves, which usually takes place on the plains. Consequently, Maasai pastoralists have to avoid the plains at this time of year. New work is focusing on vaccines for cattle. The conservation consequences are mixed: protected cattle may remove the political demand to build fences on the plains to keep wildebeest out. On the other hand, it encourages pastoralists to encroach even further into the park, which is already a major concern, as I discuss above.[17]

Interfering with Migration (Subprinciple 3)

Understanding the mechanics of migration (**subprinciple 3**) has an important consequence for conservation. If migrant populations are ever prevented from moving, they are automatically overstocked (because they no longer have access to the extra food). They will eat out their now restricted food, and the population will collapse. This has been observed wherever migrant populations have been blocked by fences, as we have seen in Botswana and elsewhere, and unfortunately as a result most migration systems of large mammals in the world are threatened with extinction.[18] When cattle migrations south of the Sahara Desert were blocked in the 1970s, pastoralists in West Africa suffered widespread starvation for the same reason.[19] The conservation conclusions from research on migration are twofold: first, if the Serengeti is to survive, migrant populations must be allowed to remain in high numbers. And second, the migrants must not be blocked by fences along the edge of the park or along roads.

Yet fences and roads are just what is being proposed. Boundaries imposed on the ecosystem, as I discuss above, limit the space resources and thus animal population sizes in general, as well as create human–wildlife conflicts. However, boundaries and major trunk roads are of even greater conservation concern with respect to migration, which underlies the working of the Serengeti ecosystem. Fences around the edge of the system are already in place north of the Mara Reserve on the Loita plains, and the first fences have (as of November 2019) been placed along the western boundary of the Ikorongo Reserve by the private owners of the Grumeti Reserves in Serengeti. There is also a proposal to build a tarmac trunk road across the middle of Serengeti. This was opposed in 2010, and the project was dropped,[20] but it has since resurfaced with all the old issues.

The proposed road cuts the northern extension of the park in two. This is the area to which the wildebeest and zebra migrate in the dry season as their refuge when food and water are in short supply. They head for the Mara River, the only permanent water supply in the ecosystem, and the region of highest rainfall, where they can find green food. Migrants remain in this region from June to November, the duration depending on the vagaries of rainfall; they usually move outside the park boundaries on both the west and east sides, just as they did a century ago, although now not so far.[21] The numbers of these animals are in the range of 1 million wildebeest, zebra, eland, and gazelle. In addition, large herds of buffalo, elephant, and topi are resident there. This vast number of animals could be feeding near the road, crossing it, and, because it is open and flat, resting there to ruminate.

There are many roads in the Serengeti National Park, but none of them produce the human–wildlife conflict that a tarmac road across the northern Serengeti is likely to create. The present roads in the park are made of gravel and are used by local traffic for tourism, by a few buses, and by local trucks. These trucks are small because all traffic currently must navigate the tortuous, steep, and narrow Ngorongoro escarpment. This road is impassable for large, 18-wheel trucks and semitrailers; indeed, they are not allowed to use the road. The traffic cannot travel at more than 50 kilometers per hour.

The problem that remains with the northern road, although currently planned as a gravel road, is that there would be a great temptation to make it part of a major trunk route across Africa, used by heavy multi-wheelers because it would be so convenient. However, there is a limit to the amount of traffic that a gravel road can sustain. The estimates of traffic use are already beyond this limit, and traffic flow will only increase in the future. So if the northern route does develop into a trunk road across the park, it will have to be tarmac, suitable for heavy traffic. The problem with such tarmac roads is not that they impede wildlife—they do not—but that they encourage fast driving. With such vast numbers of animals on the road, there are bound to be accidents, just as have occurred in Mikumi National Park, Tanzania, at Banff National Park in Canada, and in Hluhluwe-iMfolosi Park, South Africa.[22]

Accidents cause human fatalities, as has occurred at Banff, and soon enough there is pressure to build fences along the road. This may not happen immediately, it could be 20 or even 50 years away, but eventually it will happen. More realistically, it will happen in the next 10 years after the road is constructed. The important point is that once the process of building a road is engaged, there is no turning back. The concern is that the road will never be taken out again; development will proceed inevitably toward a fence. The fence will prevent the migrants from reaching their dry-season refuge containing their last water and food reserves, and so there will be a collapse in numbers and eventually the whole Serengeti ecosystem.[23] For example, the migratory wildebeest in Botswana collapsed to 10 percent of their former population when they were cut off by fences from their dry-season water and food supplies in the 1980s.[24] The general experience of fences in southern Africa shows that there are negative effects on wild species, natural communities, and entire ecosystems; indeed, they even exacerbate human–wildlife conflict.[25] Fences have already been shown to cause the end of migrating systems not only in Africa but around the world.[26] Without the migrants, the Serengeti will change into a different system. The Serengeti as the locus of a great migration will be lost.

In addition, a major trunk road attracts settlement along it, bringing buildings to the very edge of the park boundary, which will further exacerbate the human–wildlife conflict. People will have hundreds of animals in their gardens and along the streets. Inevitably, fences will be constructed along park boundaries. The drawback with fences, of course, is that wildebeest and zebra do not know what they are. They have never met them in their lives and are not adapted to tolerate them. They usually do not see the wires, or they think they can push through the fence. Fences built to withstand the charging masses will result in a catastrophic mortality with bodies lined up along it, as now seen north of the Mara Reserve. Lions will use the fence to trap prey, as they have with the fence around Keekorok Airport in the Mara Reserve. On the Loita plains, fences around the wheat farms result in many wildebeest deaths. In sum, impacts from roads and corridors include habitat loss, intrusion of edge effects in natural areas, isolation of populations, barrier effects, road mortality, and increased human access, including all its attendant problems. It is thus not hard to predict that a major human–wildlife conflict would develop where none now exists if a tarmac road were to proceed. The fundamental conservation concern is that the ecosystem would collapse.

Biodiversity (Subprinciple 4)

Social behavior contributes to the stability of the food chain and to biodiversity (**subprinciple 4**). An example of one aspect of this stability is lion social behavior where a pride of females has resident males. If males are removed through fights, new ones come in and, after killing the babies belonging to the previous males, start breeding again. The remarkable studies of Brian Bertram and Craig Packer over 40 years show that this infanticide is the normal process that takes place every few years.[27] Lions are hunted outside the national-park boundaries, in Maswa Game Reserve and Loliondo, and it is often the pride males that are killed. New males, from the pool of young bachelors inside the park, come in to replace the animals that have been shot, which counterintuitively benefits the park prides by reducing the conflicts and hence infanticide.[28] The conservation concern is outside the park; if pride males are killed too often, infanticide becomes excessive, and the pride cannot produce enough cubs to be stable. Eventually the pride disappears, thus upsetting the trophic cascade and leading to instability in the community.[29]

Long-Term Climate Change (Subprinciple 6)

One of the most important consequences for conservation concerns human-caused climate change and relates to our fifth and sixth subprinciples. One effect of climate change is the increased frequency of extreme events, in Serengeti's case droughts and floods. As we saw in chapter 8, disturbances due to weather events can be compensated for by regulatory processes (**subprinciple 5**), but only up to a point—if events become too frequent, they may overcome the stabilizing processes, and the system could change by losing species. The disappearance of savanna in the 1920–70 period caused by too many severe fires is an example.

The second aspect of climate change is its contribution to the continually changing baseline as the climate becomes warmer. All over the world, ecosystems change all the time; that is the conclusion from **subprinciple 6**. This continuous change has profound conservation consequences because it cannot be accommodated by the primary conservation strategy of legally demarcated protected areas (this includes any of the variety of legally recognized prescriptions). If an ecosystem is always changing in response to changing environmental conditions, the community of species will move accordingly. So it is quite possible that the boundaries of a protected area originally set up to contain a community may not in the future be in the right place. This is especially relevant to the movements of migrating animals such as the wildebeest. We already know that this migration extended much farther west and east of its present distribution some 100 years ago.

Human-caused climate change will contribute to this changing ecosystem. The dynamics of the Serengeti ecosystem are fundamentally based on the gradient of rainfall, with the wettest areas in the northwest *Terminalia* woodlands and the driest on the southeast short-grass plains. This means that if the environment becomes drier, the entire Serengeti community of species will move toward the northwest and so outside of the park in order to keep to their accustomed climate. But it is not just the migrants that will move; the distributions of resident birds, rodents, butterflies, and other ungulates will all move outside the protected area too.

Of course, we should expect that at the dry end, species originally outside the park will move in, and indeed we have found just this happening with the bird community around Olduvai Gorge, Lake Lagarja, and the eastern woodlands.[30] But the problem concerns the wetter end of the gradient because outside the park boundaries lies intensive agriculture, no place for most of these Serengeti species to live. Trees are largely absent, native grasses are replaced by monocultures of domestic crops, and nonnative shrubs become abundant along hedgerows. Studies of the bird communities have shown that many species, perhaps half of them, cannot live in these agricultural areas, and they disappear.[31] We already see that as rare habitats such as the Mara River forests disappear, so do some bird species such as the hornbills and bee-eaters because those habitats no longer exist outside the park.[32]

Not all species, however, respond to agriculture in the same way. We now know that specialist species, those that cannot tolerate large habitat changes or conversions such as agriculture, are more affected than more resilient species.[33] Birds that occur in many habitats—these are generalists—are found throughout Africa and are likely to be more adaptable to habitat change than those that are local endemics or specialists, living in a restricted range of conditions and habitats. This is indeed what we find for both birds and butterflies in Serengeti: where agriculture occurs, we lose the rare species.[34]

The effects of habitat modification (in this case, agriculture) tend to become more extreme up the food chain—they are magnified. This means that predators are more affected by agriculture than plant feeders, and, again, we have seen this occurring in the agricultural areas outside Serengeti. Obviously, there are many fewer large carnivores but also fewer small carnivores, and the same is seen with birds of prey. The top levels of the food chain are lost, which causes instability in the trophic cascade evidenced by outbreaks of crop pests—rodents and insects, as we saw for subprinciple 2.[35]

In summary, then, continuous environmental change, which includes human-caused climate change, will push the ecosystem outside the static boundaries of protected areas and into human-disturbed habitats that cannot support the community of native species. Thus, **subprinciple 6** leads to perhaps the most alarming of all the conservation consequences because it implies that none of the currently protected ecosystems in the world will be within their boundaries a century or so from now.

Alternative Stable States (Subprinciple 7)

Slow, continuous change, however, can reach a point where it becomes too severe for regulation to redress the system, and a new species complex develops relatively rapidly—what we term a change in state, our **subprinciple 7**. We explored this situation in chapter 10, where we showed that the savanna can exist as two states, one with mature trees and elephants and a second one without mature trees (only small seedlings) and the same density of elephants. The change from woodland to grassland was caused by excessive modification due to burning. The return to a woodland state required not just a reduction of burning but also a reduction of elephant browsing. These two states—savanna and grassland—are perhaps normal in drier parts of Africa, and we see from the pollen work in Tsavo National Park that woodland and grassland have alternated every few hundred years.[36]

However, the conservation concern arises when too much habitat modification pushes the system into an unwanted state, one that cannot be reversed. Because the Mara River forests are special, we don't want to lose them—they support many animals not found elsewhere in the system, including some endemic species. We described in chapter 10 the unstoppable unraveling of these forests into savanna as a result of the opening of the canopy by burning. Fruit-eating birds necessary for germination of forest trees disappear (especially hornbills and barbets), germination collapses, and the forest disappears over a period of 40 years. At this stage it cannot recover.

A less obvious example is the disturbance to the soils of the short-grass plains by the uncontrolled creation of vehicle tracks, mentioned above. Most of these tracks, although now unused, are still visible even 50 years after the disturbance—and they support a different plant community. Trails caused by centuries of migrating animals are evident across the short-grass plains, as are holes made by spring hares and other species, but these are all natural disturbances for the system. The extensive parallel vehicle tracks are quite unnatural but are stuck in this disturbed state—one that we need to avoid.

————

I have outlined in this chapter the most important conservation consequences derived from the scientific findings on how the Serengeti ecosystem works. In particular, the loss of biodiversity causes instability; interference with migration will completely alter the nature of the Serengeti ecosystem; the continually changing environment forces the community to move outside established boundaries; and excessive modification is causing some habitats to be irrevocably destroyed. To address these changes, fixed boundaries must become flexible. This could be achieved by making areas outside the present boundaries more nature-friendly—for example, by regrowing more suitable habitats within agricultural areas so that wild species can live there and tolerating wild ungulates alongside domestic livestock (this is known as *community-based conservation*). Alternatively—or in addition—a system of new reserves connected with the Serengeti can be set up as contingencies—as security—to be available when the system changes and animals move into them.

These are the lessons for the future of Serengeti. Now we will see how these variations of regulation, the subprinciples, apply to other ecosystems around the world.

13

Lessons from the Serengeti

So far, we have identified seven variations—the **subprinciples**—of the **principle of regulation** that explain how ecosystems work. There are certain to be more, having to do with the flow of energy and nutrients, the process of pollination, and others, but we do not have sufficient information for Serengeti to elaborate on them here. However, these seven subprinciples are sufficient to move the argument forward.

All around the world we see evidence of ecosystems that one way or another are in distress, resulting in what has recently been labeled by Jim Estes and colleagues as the *downgrading* of ecosystems. Repairing of these damaged systems has been called *upgrading* by Rob Pringle of Princeton University.[1] It turns out that many if not all problems relating to downgrading stem from an abuse of one of these subprinciples. We can start by examining some of these downgraded ecosystems around the world.

Interfering with the Regulatory Process in Populations (the Fundamental Principle of Regulation)

Our fundamental principle is that systems can be regulated. Interfering with the regulatory process can cause distortions of the species community and other unwanted outcomes. In the early 1960s, scientists in the Kruger National Park, South Africa, recommended that elephant and buffalo populations be culled because these species would become too numerous for their habitat and cause ecosystem collapse. The concept of regulation was never considered. Consequently, a major disturbance in the form of industrial killing took place for the better part of 40 years. Elephant and buffalo numbers were held down, so that regulation was prevented and the effects of the populations on the vegetation were unnatural, artifacts of the culling. A similar argument supported the culling of elephants in Murchison Falls National Park, in Uganda, and of hippo in Queen Elizabeth National Park, also in Uganda and also in the 1960s.[2] Subsequently in Kruger Park, culling has been reduced or even stopped; the populations have not gone through the roof as feared, and regulation has been detected.[3]

So far, we have considered that regulation is relaxed naturally when populations get very small because there is plenty of food and habitat. However, there is a twist to this process if populations get too small. It was first noticed in the 1930s by an American scientist, W. C. Allee, who proposed that if individuals were so few on the ground that they may never meet to mate, reproduction would decrease instead of increase and the population would collapse to extinction. This has since been called the *Allee effect* (see chapter 11).[4] Failing to find mates is not the only process that can distort the regulatory process in very low populations. We have already seen (chapter 5) that predators can drive populations to extinction if they are secondary prey and are already at a low enough level. Small populations can also go extinct simply as a result of chance events; the reproduction of the few living animals cannot compensate for the deaths due to accident (see chapter 11 discussion of the North Atlantic right whale and the Serengeti black rhino). The Iranian Cheetah Society reports that there may be less than 40 Asian cheetah remaining in Iran; mortality exceeds births, and many do not breed at all. In essence, the principle of regulation becomes distorted when populations fall unnaturally low through habitat loss, disappearance of food, or too much disturbance from humans, mesopredators, or climate change.[5] There are many examples of this Allee effect, and as species populations are declining rapidly all over the world, many of them are collapsing into this "extinction trap."

Interfering with Bottom-Up Processes
(Subprinciple 1)

All ecosystems depend on nutrients, energy, water supply, and protection against pollution and unwanted foreign species. Nutrients originate at the bottom of the food chain in soils and plants on land, and, in water, through small plants and animals that live in seas and lakes.[6] If we alter the flow of any of these important ecosystem supplies, the whole food chain is affected from the bottom up, often with the loss of predators and the appearance of species that should not be there.

Whenever agriculture applies too much nitrogen and phosphorus fertilizer to arable land, a lot of it is washed into lakes and rivers. This pulse of nutrients changes the bottom of the food chain from bacteria to a strange group called cyanobacteria (blue-green algae). A normal lake food chain has bacteria, small algae, small animals (zooplankton) that feed on them, small fishes that feed on zooplankton, and larger, top fishes that feed on the smaller fishes. But when too much phosphorus relative to nitrogen enters the lake, the blue-green algae thrive, taking all the oxygen, and the rest of the food chain dies from lack of oxygen. This is the process called *eutrophication*, and it kills lakes.

Interfering with the plant level will, of course, alter the food chain above it. Overgrazing by cattle in the areas north of the Mara Reserve, Kenya (part of the Serengeti ecosystem), has since the 1990s altered the balance of competition between grasses and flowering plants to the disadvantage of the grasses. Consequently, thickets of shrubs and trees have developed, excluding the grasses and causing a collapse of populations of wild grazers such as topi, kongoni, and wildebeest. The same interference in the 2020s is taking place in the eastern woodlands of Serengeti due to incursions of cattle from outside the park.

Interference in the form of supplying extra food to snow geese on winter feeding grounds in the southern United States has resulted in much higher populations feeding on the Canadian arctic tundra around Hudson Bay, causing damage to their summer feeding grounds. Geese have been forced to move to other areas, and the ecosystem has changed.[7]

Loss of habitat is equivalent to the loss of an essential resource for all species—a place to live in. This is really a bottom-up distortion resulting in declining populations, sometimes to extinction. Lion numbers in Africa, cheetah numbers in Africa and Arabia, and countless numbers of bird, mammal, and insect species have declined because of loss of habitat. The red-cockaded woodpecker in the southeastern United States, which requires open longleaf-pine forest, has declined because the pine habitat, which depends on frequent burning, has declined.[8] The orange-bellied parrot in Australia, one of the most threatened parrot species in the world, is close to extinction due to habitat loss followed by Allee effects.[9] These are just two of a long list.

Loss of animal prey causes problems for predators further up the food chain. The rinderpest virus of 1890 killed off so many wildlife species that lions turned to eating people because they had no other food source. In what became Tsavo National Park 60 years later, all the buffalo died, and lions began eating the Indian laborers building the Mombasa–Nairobi railway. These lions were the famous "Maneaters of Tsavo." Attacks also occurred in Uganda, and just west of Serengeti. Indeed, such attacks contributed to the evacuation of agricultural land on the southwest borders of Serengeti at the beginning of the twentieth century.[10] In the 1970s, in southern Tanzania, villagers killed off all the wildlife that lions were feeding on, and lions then killed the villagers.[11]

Interfering with Top-Down Processes (Subprinciple 2)

One of the most frequent underlying causes of ecosystem collapse is the interference with top-down interactions in the trophic cascade (see chapter 5).[12] Usually this occurs through the loss of top predators, but on islands it is often the appearance of unwanted top predators that causes the problem. In many parts of the world top predators have disappeared as human populations and agriculture have spread, particularly in Europe and parts of North America. In North America, loss of top predators has resulted in outbreaks of white-tailed deer and excessive browsing of trees.[13] In Europe, wolves and brown bears that once regulated moose and red deer have been extirpated from most of the continent. Their absence has led to an overabundance of herbivores, which has reduced the survival and natural regeneration of trees. The disappearance of the lynx that control the smaller roe deer has also resulted in extreme browsing.[14] Almost everywhere in Europe, regulation of herbivores by predators is prevented by a landscape highly modified by agriculture, urban development, and human hunting.[15]

In marine systems, loss of large fish species through overfishing has caused major changes in the trophic cascade,[16] as well as collapse of coastal ecosystems.[17] For example, the salt marshes off the New England coast of the United States support spartina grass, which holds the mud banks together. Recently these spartina banks have died, leaving unvegetated, eroding mudflats, and locking the marsh into an unwanted sterile condition. The die-off was caused by overgrazing from an outbreak of the herbivorous purple marsh crab. This crab was released from top-down predator regulation by large fish and other predatory crabs that had been overharvested by humans.[18]

But it is not just the herbivores that increase with the loss of top predators. Often there are smaller predators, normally regulated from the top, that increase hugely when top predators disappear, and cause severe problems lower in the food chain. Coyotes increased with the loss of wolves, as seen in Yellowstone National Park;[19] baboons increased when leopards were removed in Ghana;[20] and there are countless examples around human habitations in Europe and North America where crows, magpies, raccoons, foxes, gray squirrels, and skunks have taken over in the absence of birds of prey and top carnivores, with the consequence that small birds cannot nest successfully.[21] This outbreak of small predators is called in the jargon *mesopredator release*—a distortion of the predator cascade.

John Terborgh recorded what happened when parts of the Amazon forest were cut off as islands in a new artificial lake in Venezuela. These islands were left without their top predators, the primates. He found that leaf-eating insects, among other species, increased to such vast numbers that they stripped the trees of their leaves, and the trees died. The forest literally collapsed as he recorded these events. Mary Power showed that when big mouth bass were removed from streams in California the abundance of smaller fish increased dramatically.[22]

Jim Estes has found that when sea otters disappeared off the coast of Alaska due to excessive hunting for the fur trade in the early 1900s, sea urchins increased and grazed down the algae to a lawn, with the result that fishes that lived in the algal (kelp) forests disappeared. The system was locked into this barren state until sea otters returned a century later. Now, however, there has been a new disturbance due to climatic and temperature changes in the Pacific Ocean dating to the late 1970s, which also appear to have long-term ecosystem consequences for the trophic cascade. The temperature shift has changed the complex of fish species in the North Pacific. This new fish community is of much lower food quality for Steller's sea lions. In the mid-1970s this sea lion population was increasing and was around 250,000. It dropped rapidly to 100,000 by 1990 and to 50,000 by 2000. Killer whales, which had previously been feeding on Steller's sea lions, switched to feeding on the recently returned sea otters, which caused a new sea urchin outbreak.[23] More recently, along the coast of California, sea stars feeding on sea urchins were largely wiped out by a disease, leading to a population explosion of sea urchins. This, in combination with warmer-than-usual ocean waters from 2014 to 2017, has decimated kelp (algal) forests by more than 90 percent along 350 kilometers of coastline.[24]

Because of the chain of events in trophic cascades, there are often far-reaching indirect consequences of the removal of top predators. The Channel Islands off the California coast are home to two coexisting and rare small carnivores, the island fox and skunk. Several decades ago, feral pigs became established and sufficiently numerous to provide prey for golden eagles, which took up residence there for the first time in 1992, and these eagles have subsequently increased in number. The eagles began to hunt island foxes as secondary prey and have driven fox numbers down. Skunks, being nocturnal and therefore less subject to predation by eagles, have increased because they no longer have competition from foxes.[25]

In many ecosystems, top-down regulation has been distorted by the introduction of foreign predators—those not supposed to be there and to which the native species are highly vulnerable. In North America, the brown-headed cowbird is a nest parasite of many small songbirds. Its population is increasing, and it has caused the decline of at least two species, Least Bell's vireo and the black-capped vireo. The ultimate cause can be traced to events that happened more than a century ago. This cowbird's original range lies in the prairies, where it used to follow bison. As more land was opened by agriculture across North America, the cowbird spread into new, previously forested areas, where it parasitizes native bird species that are not adapted to tolerate it. The appearance of a new top predator from a distant ecosystem has changed the trophic cascade.[26]

Most of the problems with unnatural (foreign) top predators occur on islands. The introduction to Australia of foxes and cats from Europe caused the extinction of many small marsupials at the turn of the twentieth century. The problem was exacerbated by the additional introduction of a foreign prey, the European rabbit, which inflated the numbers of foxes and cats, making their effects all the more disastrous. There are numerous examples from small oceanic islands where introduced species caused havoc: cats and rats on Macquarie Island in the Subantarctic; goats, pigs, rats, and mice on Lord Howe Island off eastern Australia; goats in the Galapagos Islands; rats on Henderson Island in the remote eastern Pacific—these are just a few of a lengthy list. These animals, high on the food chain, have all caused serious declines or extinctions of the native prey. Great efforts are now being made to remove these unwanted members of the food chain.

A particularly complex example of an unnatural predator occurred in New Zealand 700 years ago, and its effects have remained to this day. Prior to the arrival of humans in New Zealand, around the year 1280, vast numbers of small burrowing petrels and shearwaters (seabirds, thought to be in the billions) roosted and nested on the forest floors of the mainland as well as outer islands. These numbers were large enough that their feces provided considerable amounts of nitrogen to the ecosystem, which at that time was more than 85 percent forest; in effect, these nutrients from the sea via the seabirds fertilized much of the land ecosystem, maintaining a diverse insect fauna, as well as numerous vertebrates such as lizards and birds dependent upon the insects. The arrival of the Pacific rat with the first humans caused a catastrophic decline of local seabirds within a century, and with them the associated invertebrates, lizards, and terrestrial birds. The moa, a large flightless bird of New Zealand, may have been hunted to extinction several hundred years ago, but it is also possible that its extinction was caused by the reduction in its food supply, which had been nourished by the nutrient flow from seabird waste. Pacific rats also selectively feed on seeds and seedlings of forest plants, so that the composition of the forest flora is now considerably different than it was before they arrived.[27] This rat became the new top predator in the thirteenth century, a new keystone species, radically changing the New Zealand forest trophic cascade and affecting both top-down and bottom-up processes.

Many ecosystem problems come from the interplay of several different factors. An ecosystem may be adapted to one form of disturbance but unable to cope when a new disturbance is added; the combination creates more mortality than reproduction can cope with. Stan Boutin, who worked with me in Serengeti, recounts that in British Columbia, mountain caribou, a beautiful subspecies that lives in high mountain forests, was the normal (primary) prey of wolves. In the early twentieth century, however, moose, not previously seen at these lower latitudes, began to spread south from Alaska as forests were opened by human logging. As moose became their primary prey, wolf numbers increased, and caribou populations crashed because there were more wolves to eat not only moose but also the caribou that had become their secondary prey. But it was not just predation that caused this; it was also the loss of the caribous' essential mature forest habitat, which reduced their numbers to sizes that could live in the small patches left by logging. It was the combination of less habitat (bottom-up) and more predation (top-down) that the caribou could not tolerate.[28]

In Western Australia, the burrowing bettong, a small (1 kilogram) marsupial, was once abundant throughout the desert regions. Adults tolerated drought, but this came at the cost of low reproduction (animals can't have both high survival and high reproduction; one has to give). In the nineteenth century, they experienced competition for food

and burrows from introduced rabbits, to which bettongs were not naturally adapted, but they survived droughts better than rabbits, which
have the opposite tendency—short survival and high reproduction.[29]
However, early in the twentieth century, when foxes and cats spread
across Australia, the combination of competition (bottom-up) and predation (top-down) exceeded the low reproductive capacity of bettongs
and many other marsupial species; these species went extinct on the
mainland. Luckily, some of the marsupial species survived on a few offshore islands that did not have foxes, so we still have some alive today.[30]

Of all the disturbances discussed in this chapter, altering the trophic
cascade is perhaps the most frequent and most important in causing
problems for humanity. The correct trophic cascade is vital for the
proper functioning and services in human ecosystems.

Preventing Migrations (Subprinciple 3)

Seasonal movements and migrations are features of most ecosystems;
in some circumstances, migration allows large populations because of
the extra food they can access and the reduction in predation (chapter 6). These two observations from John Fryxell's work predict that if
a migratory population is allowed to fall too low, it can become trapped
at a low level, held there by reduced food supply and increased predation, and may never climb back out again, an aspect we will return to
later when we look at alternative stable states.

Here we consider the prevention of movements and the loss of habitat and food. The placement of artificial water holes in Hwange Park
(Zimbabwe), Kruger Park (South Africa), and the Sahel of northern
Africa at first seemed a sensible strategy to benefit particular ungulate
species searching for water. But when presented with permanent water
supplies, species that normally had migrated to find seasonal watering
places became resident. This led to severe overgrazing and damage to
the vegetation in the habitats with the artificial water holes. In Zimbabwe, the natural regime in the 1930s consisted of broad-leaved woodland
(called miombo) and temporary water holes on ridgetops used by elephants in the wet season. In the dry season, the elephants migrated to

the valleys, where they had permanent water in the rivers and rich riverine grass. However, humans fenced off the valleys for their own use, stopping the migration of elephants and leaving them to live in the miombo. Since there were no permanent water supplies in the miombo, artificial ones were provided. Elephants were now compelled to live year round in a habitat that was not adapted for constant browsing. At the same time, the number of elephants that could be supported by this very low nutrient vegetation was much lower than previously, when valley grasslands were available. Consequently, the woodland was damaged in large halos around these unnatural water holes. Managers were then obliged to kill elephants in order to rectify a situation that had been initiated by humans through preventing the annual migration.[31]

In Kruger Park, in the 1930s, there was a migration of wildebeest to the higher-rainfall areas west of the park in the dry season. Then ranchers claimed these rich, wet areas, and fences were constructed to keep wildebeest out. This prevented the migration, and wildebeest numbers declined to a level suitable for a sedentary population, again using artificial water holes.[32] A similar migration of wildebeest occurred in Botswana; they moved from good feeding areas in the wet season to the great Okavango Swamp in the dry season. Fences were put up to separate wildlife from cattle in order to prevent the spread of foot-and-mouth disease. This killed the migration, and the wildebeest population has collapsed to almost nothing.[33]

The Sahel of Africa is a band of semiarid vegetation just south of the Sahara Desert. In the (short) wet season, the grasses are very nutritious, and cattle herders traditionally moved their animals north to the Sahel to fatten up and breed. When the season ended, they moved south again, sometimes long distances, to water holes that were permanent but where the food was poor. Just as with Serengeti wildebeest, the ability to access the temporary good food allowed high numbers of livestock. Then well-meaning but misguided government officials drilled more water holes and prevented the pastoralists from moving (often they moved into another country, which governments did not like). As a result, there was overgrazing around the water holes triggered by drought, resulting in mass starvation of both cattle and people in the 1970s and '80s.[34]

The overwhelming evidence provided by these three examples shows that the blocking of migration, our third subprinciple, has resulted in unwanted ecosystem consequences.[35]

Loss of Biodiversity (Subprinciple 4)

Intact eucalypt forests in Australia support many species of coexisting, endemic birds called honeyeaters. In eastern Australia, this forest has been opened up since the early 1900s for logging, persistent livestock grazing, and agriculture. One honeyeater, the large noisy miner, benefited from this open habitat. It now dominates the bird community by aggressively excluding most of the smaller honeyeaters that formerly fed on insects and prevented outbreaks, particularly of plant-sucking insects called psyllids. Without the small honeyeaters, trees in the open suffer considerable damage from these insects and eventually die—a consequence of loss of biodiversity.[36]

Reef-building corals (*Monastraea* species) host dynamic, multispecies communities of tiny, one-celled organisms called dinoflagellates (*Symbiodinium* species). As a consequence of abnormally high irradiance and seawater temperature, corals exhibit bleaching, a loss of photosynthetic ability. But corals with a more diverse community of symbiont dinoflagellates are better able to withstand these stresses and exhibit less bleaching, evidence that biodiversity protects the reef from disturbance.[37]

In any ecosystem, loss of keystone species results in a loss of biodiversity at other levels in the food chain. As we have already seen, wildebeest in the Serengeti determine the abundance and diversity of the large-mammal community by affecting competition and predation and through complex indirect effects. Wildebeest also influence the biodiversity of other, less obvious components of the system: they determine both the physical structure and the species composition of grasses and herbs of the short-grass plains, notably allowing a large number of small flowering-herb species. The bird community is affected by grazing, which causes changes in grass height that determine nesting sites, mode of feeding, and food availability. Of the eight commonest bird species in Serengeti feeding in the grass layer of grazed sites, only one became more abundant in ungrazed sites, whereas all others were reduced by 50 to 80 percent.[38] Thus, the loss of wildebeest a century ago would have reduced biodiversity and changed the way the ecosystem worked. The conservation consequences of the loss of wildebeest as a keystone species would be disproportionately greater than those due to the loss of black rhino or wild dog, both of which have occurred with seemingly little impact on the system.

In general, the greater the number of species in an ecosystem, and the greater the number of connections among species through competition and predation, the greater the stability of that system.[39] The loss of biodiversity, often but not always through loss of a keystone species, results in outbreaks of some species and the loss of others, causing problems for both natural and human-use areas, as was predicted by subprinciple 4.

Too Much Disturbance (Subprinciple 5)

The fifth subprinciple holds that all ecosystems are disturbed to some extent by variations in the environment, but that these disturbances can be countered by regulatory mechanisms. However, when there is too great a disturbance—or not enough disturbance—the ecosystem can change. But this depends on the type of species involved: species that are wide ranging and live in a variety of habitats can tolerate greater or more frequent disturbances. These species are called generalists and are said to be *resilient*. Conversely, species that have small ranges and narrow ecological niches are specialists and are said to be *fragile*. A moderate amount of hurricanes, floods, droughts, and fires can be tolerated and can actually help to shape the ecosystem.[40] Moderate degrees of burning in savannas can reverse vegetation changes—called *succession*—and hold a system at an intermediate state called a *fire disclimax*.[41] Repeated disturbances such as hurricanes over the past 14,000 years are thought to have structured some tropical forests and islands in the Caribbean.[42]

But disturbances that are too frequent or too extreme can radically alter an ecosystem, changing it to a different state. If the ecosystem does not return to its previous state when the disturbance is lifted, we say that it is in an alternative stable state (our seventh subprinciple), of which we will see examples later (referred to as unwanted alternative stable states).

On Macquarie Island in the Subantarctic, for instance, there has been widespread dieback of cushion plants and bryophytes in alpine habitats, especially the keystone cushion plant *Azorella macquariensis*, to the point that it has become critically endangered. This plant is adapted to a persistent cold, wet, and misty climate. The dieback was caused by an increase in the frequency of summer droughts due to warming climates, an increase in a disturbance to which the plant is not adapted.[43]

Too little disturbance, as noted, can also cause ecosystem problems. Reduction of fire in itself has produced disastrous consequences in the conifer forests of British Columbia. Frequent fires, with their patchwork of burns, leave only small areas with a ground layer of dead, combustible material suitable for fires. By contrast, fire prevention led to catastrophic fires in 2017 and 2018 over large areas. Too little disturbance results in later conflagrations.

Accommodating Continuous Ecosystem Change (Subprinciple 6)

Ecosystems are dynamic; they change over time (subprinciple 6). Constraining this process by putting static boundaries around them leads to problems. The Atlantic mountain ranges of eastern Brazil are valuable biodiversity hot spots and a focus for conservation. In the southern highlands, there is a unique vegetation mosaic of *Araucaria* forest and campos, a diverse and special grassland. The *Araucaria* is itself a strange forest, perhaps better known for one of its species, the garden favorite called the monkey-puzzle tree. This vegetation mosaic is protected within the boundaries of parks such as the Aparados da Serra National Park. The conservation threats here are severe—the campos is being displaced by forest, while the forest is itself being destroyed by logging, fire, and grazing. Reduction of these disturbances may maintain the forest but not the campos. However, evidence from fossil pollen and microfossil charcoal reveals a history of continuous and radical changes to the vegetation over the past 11,000 years. Between 11,000 and 4,000 years ago, the current protected area was largely grassland. Then, from 4,000 to 1,000 years ago, the *Araucaria* forest expanded into the region, forming a network of forests along streams, while grassland vegetation still dominated. Forest then expanded further, reducing the campos to present conditions from about 430 years ago because of higher rainfall and a shorter annual dry season. The evidence shows that this rare and diverse vegetation mosaic exists in a dynamic and continuously changing state, determined by climate. Only a few hundred years ago, the forest barely existed in the current protected areas, and it is likely that in future the forest will disappear from these areas again but reappear in other areas that are currently not protected. Interfering with this change using static boundaries and manipulations of disturbances will only increase the mosaic's vulnerability. Boundaries need to be as flexible as the vegetation.[44]

A similar case is seen with the range of the giant panda in China. The pandas feed on a restricted diet of a few bamboo species, in particular of the genus *Fargesia*. Several of these species flower synchronously over large areas every 70 years or so—there was such a flowering in the mid-1970s. After flowering, the plants die and large areas of lower-elevation bamboo, required by pandas for most of the year, are no longer available to them. The current range of pandas is now so restricted by static boundaries that they have become vulnerable to this large-scale die-off of their food plants. This is because panda conservation areas in the 1970s were not sufficiently spread over their historical range to include territory outside the area of bamboo die-off, severely limiting the panda's ability to recover from natural fluctuations in its food supply. Redesigned reserve networks with corridors connecting isolated patches have allowed the flexibility necessary for animals to search a wider area.[45] In general, the static boundaries of protected areas will not be able to contain the species they were originally designed for because the environment and habitats will change. As a result, we are already seeing the disappearance of species from protected areas.[46] I have already raised this issue for Serengeti in chapter 12.

The conclusion that continuous change must be accommodated rather than resisted (subprinciple 6) applies to many aspects of global climate change, including, for example, the changes in sea level that will flood huge areas of the Sundarban Forest in the Ganges-Brahmaputra River delta of Bangladesh, and the melting of permafrost on which several towns in Siberia are built. Human-caused climate change is now the most visible manifestation of continually changing ecosystems, altering community composition by shifting species ranges at different rates toward the poles, away from the tropics, and higher up mountains, especially in the warmer latitudes.[47] For example, in Europe, birds must move their breeding locations to match the earlier timing of insect-food hatching because of warmer spring temperatures.[48]

Unwanted Alternative Stable States (Subprinciple 7)

Disturbance to the complex of species in a community can lead to a distortion of ecosystem processes such as the water flow in rocks and soils. We have already mentioned several examples above. Another example is Australia before European settlement. The continent at one time was largely covered in eucalypt woodland. In the past century, agriculture replaced nearly all of this woodland, especially in Western Australia. Eucalypts kept groundwater levels down through evaporation of water from their leaves (transpiration processes). When the trees were removed, groundwater levels rose, water evaporated at the soil surface, and salt deposits made the soil unsuitable not only for crops but also for native plants. Now large areas of Australia have a major problem with salinization of soil and groundwater upwelling, with a resultant decline in agricultural productivity. In response to this ecological (and economic) problem, Australia has had to adopt the expensive policy of revegetation.[49] Ecologists would say that the eucalypt ecosystem has become locked into an unwanted saline state because of excessive disturbance to the woodland.

Alteration of soils is a common feature involved in changes of state. The high Arctic tundra is a slow-reacting ecosystem, but it is highly sensitive to disturbance. Michael Becker and Wayne Pollard measured the recovery of the tundra after the construction of an airstrip at an old weather station in 1947. They found that, 60 years later, the disturbed vegetation and sensitive, ice-rich permafrost had not recovered relative to undisturbed tundra and instead showed a progression toward a new plant composition and a different stable state.[50]

Deforestation of the Amazon for crops and cattle grazing has resulted in a loss of nutrients in the sandy soil that prevents forests from regrowing, so that a treeless state prevails in the disturbed areas. Overgrazing is particularly evident in many areas. Because of the heavy grazing pressure, grasses die off and are replaced by inedible species. In another part of the world, the semiarid regions of the Negev-Sinai, in Israel and Egypt, and the Sahel of Africa, a shrub and herb layer acts as a blanket on the soil, retaining moisture and heat overnight. During the day, thermal up-currents carry moisture from both the soil and transpiring plants to upper levels, where it condenses as rain. This supplies the plants and soil with water, completing a self-sustaining (positive feedback) system when in an undisturbed state. In contrast, overgrazing by livestock leaves a bare soil surface, allowing higher reflectance of sunlight off the surface and greater cooling at night. There are fewer up-currents, and these carry less moisture. Consequently, overgrazed areas have much less rain and become even more parched; this is also a positive feedback.[51] The vegetated state switches to the denuded one through the disturbance of overgrazing; stated differently, there is a threshold level of disturbance (grazing) above which one state switches to another. A similar switch in vegetation state occurs in Niger, West Africa, where overgrazing has altered the vegetation structure, leading to reduced water retention, increased soil loss, and further vegetation loss. The system is now locked into this arid, infertile state.[52] Quite probably, this has been the process in the Fertile Crescent of the Middle East, which, as the name implies, several thousands of years ago was the breadbasket of Western civilization but is now largely exposed soils and rock after centuries of overgrazing.

Livestock have grazed the grasslands of Inner Mongolia for centuries. However, between 1948 and 1967, livestock numbers increased steadily, and the cattle gradually grazed down the steppe almost to a lawn. This is just the grass height that Brandt's vole[53] prefers, and this rodent suddenly increased to high abundance through much more frequent outbreaks of its numbers. Similar changes in the state of grasslands are occurring throughout central Asia, particularly on the Tibetan Plateau.[54] In Inner Mongolia, Brandt's voles now maintain an even lower grass height, where they outcompete the livestock so that the steppe is in a new, unwanted state. To move out of this state will require both the removal of voles and a reduction of livestock grazing, so that the grassland can regrow to a height that voles avoid. One social consequence of this unwanted change in ecosystem state, which makes itself felt far from the original disturbance, is that Beijing, hundreds of kilometers distant, now experiences much more frequent dust storms.[55]

Moderate levels of herbivore disturbance reduce the effect of dominant competitors, as we saw with the keystone species of the intertidal zone (chapter 7), thus increasing species diversity.[56] Grazing by large herbivores is a form of top-down restructuring of the plant community. If this grazing becomes too severe or persistent, however, the plant community may revert to an alternative state such as denudation of the soil (as mentioned earlier for semiarid areas), or may devolve from grassland to woodland, as in savanna areas.[57] Similarly, agriculture is a top-down effect of humans on plant communities. If such changes are too extreme, foreign species can invade, taking over and creating new states.[58] In the forests of the Annamite range of Laos, in southeast Asia, shifting agriculture has been the norm, probably for centuries: patches of forest are clear cut, burned to release nutrients, planted with crops for two to three years, then left to regrow for two to three decades while the villagers move on to new areas. But some villagers have now become sedentary, and the resultant more frequent cutting over many years has so depleted soil nutrients that the invasive grass *Imperata cylindrica* has taken over entire mountainsides. The forest cannot regenerate, these areas are no longer productive for humans, and rodent outbreaks are now frequent (another example of an unwanted stable state).[59]

A combination of overharvesting and environmental change can cause fish stocks to collapse and has resulted in alternative stable states. A particularly clear case occurred on Georges Bank, a shallow offshore plateau on the East Coast of the United States. In the early 1990s, the benthic-fish community (bottom dwellers) switched suddenly, over an extremely large area, to a pelagic community (everything above benthic) as a result of overharvesting, especially of Atlantic cod (*Gadus morhua*). This change in ecosystem state has been exacerbated by decadal variability in water temperature in the northwest Atlantic.[60] In the Caribbean, overharvesting of fishes has caused catastrophes and large-scale degradation of coral reefs, resulting in a new state.[61] It is almost certain that the extinction of mammoths and other giant species 12,000 years ago was a result of the overwhelming combination of human interference (hunting, habitat alteration, and disease) and rapid climate change at the end of the Ice Age. Most species can tolerate perhaps one disturbance, but a combination overwhelms their abilities to reproduce and adapt.[62]

There are now several examples of a shift to a new state caused by interference with migration. Prior to the Second World War, there was a large migratory caribou herd in the Yukon, called the Fortymile herd. The herd, formerly numbering in the hundreds of thousands, was heavily persecuted during the Second World War. Animals were shot by army personnel for food, and numbers declined to only about 10,000, possibly less than 5 percent of their original number. Wolves are now the main cause of death in this herd, and although there has been much effort to restore the population, it was stuck at 14,000 animals for almost 60 years. Since 2014, with some control of wolf numbers, predation has been reduced and caribou numbers have increased again to about 50,000.[63] It is possible that the same situation applies to the saiga antelope on the Asian steppes. These were once a million or so in number but have declined precipitously from disease and overhunting in the past 20 years, while wolf numbers have increased considerably.[64]

Global climate change may also create unwanted states by increasing the fragility and decreasing the resilience of human-stressed ecosystems. An increase in extreme weather (stronger storms, longer droughts) can push the adaptations of species beyond their capacities—especially the fragile species—resulting in a change in ecosystem state.[65] One example is the possibly irreversible set of changes in the ecosystem of the Southern Ocean encircling Antarctica. Reductions in the extent of sea ice in winter, which may be driven by climate change, are affecting the bacteria and tiny algae that krill—small shrimp-like creatures—feed upon on the underside of the ice in winter. The abundance of krill has declined drastically, and this in turn limits the recovery of whale populations, which depend on krill.[66] The Great Barrier Reef, off eastern Australia, is experiencing increased frequency of coral bleaching precipitated by warming seas and more hurricanes. These reefs have been around for tens of millions of years and are likely to have seen such changes in the past. But when human interference is added to natural changes through the runoff of excessive nitrogen fertilizer, herbicides, and pesticides from the sugarcane fields along the Queensland coast, the reefs are overwhelmed and massive mortality ensues. The combination of factors makes the reef community more sensitive to both climate change and human abuse.[67]

As these examples illustrate, ecosystems have a threshold for change, and if disturbed too much, they may move into an unwanted state (subprinciple 7).

Upgrading

In this chapter I have shown how problems with human and natural ecosystems can be understood in the context of the seven variations (subprinciples) of the principle of ecosystem regulation. Understanding is a first step toward finding solutions to the problems by developing management and conservation strategies. However, there is some indication that, given a bit of help, ecosystems can repair themselves, a process called *upgrading*. The Serengeti itself is an example of such upgrading. The system had been severely changed by the 70-year disturbance of rinderpest (1890–1960). After the disappearance of the disease in 1963, we saw remarkable rebounds in almost all components of the community. Wildebeest and buffalo both increased fivefold, a huge change for such large animals. Their grazing reduced another disturbance, fire, so that trees were able to regenerate for the first time since the 1920s. These trees, particularly the stink-bark acacia, are favorite foods of elephant and so contributed to their rebound in populations from another disturbance, the ivory trade of 1840–1880 (chapter 4). With the trees came an increase in the bird species that nest and use them, such as hornbills, rollers, and turacos. The increased grazing on the plains changed the grasslands and promoted bird communities that live in short-grass swards. The higher numbers of wildebeest increased the top-predator populations of lions and hyenas. But some species also decreased with the changing ecosystem: wild dogs effectively disappeared when previously they occurred over the entire savanna. Herbivores such as ostrich also declined—flocks of 200 or more were recorded from the 1890s up to the early 1960s, numbers that are not seen today—as did roan antelope, which also occurred throughout the savanna. The latter are now reduced to one small herd in the west. So some species have increased and some have decreased in abundance as the system has repaired itself since the disturbances were removed; even today, Serengeti is still rebounding from rinderpest.

Gorongosa National Park, in Mozambique, was in the 1960s a complete and well-functioning natural ecosystem in the miombo woodlands of southern Africa. A 15-year civil war (1977–1992) removed most of the large mammals from the system. Elephants were reduced to a few dozen, lions to one pride, and buffalo vanished altogether, along with many other species. Meanwhile, waterbuck grew numerous. This severely disturbed ecosystem is now recovering slowly, both naturally and with help from managers who are reintroducing species, a classic case of upgrading (see chapter 14). The ecosystem is now being monitored by Rob Pringle of Princeton University and his colleagues to see how it recovers and whether it returns to the pre-disturbance community.[68]

In the Kalahari region of southern Africa, once-stable sand dunes have become active again as a result of human disturbances such as grazing and farming. Active dunes lead to wind transport of sands, which can negatively affect other areas far from them. It is now known that if the disturbances persist, the dunes become locked in a new stable state of movement. However, if the disturbance is removed early enough, the dunes can recover their long-lived (perennial) grasses, and stability returns; a natural upgrading can occur.[69]

Wolves were extirpated from Banff National Park, in Canada, many decades ago. Elk grew numerous and used the entire Bow Valley system in the absence of wolves. In 1986, wolves returned by natural self-reintroduction, and the elk population responded by using human-dominated areas near the town of Banff, where they felt safe, and avoided areas frequented by wolves. By comparing these two areas, Mark Hebblewhite and colleagues showed that a new trophic cascade formed naturally; where there were wolves, elk density was lower and survival of adults and calves was lower. Consequently, browsing of willow and aspen declined and more plants survived, and because of this more beaver lodges appeared and bird diversity and abundance along streams increased. Beavers and birds illustrate the indirect connections in the trophic cascade, which, as we see, can reassert itself in a natural upgrading, if opportunities permit.[70]

———

We have presented this litany of examples from around the world to show that ecosystem problems can be understood in the context of the subprinciples that explain Serengeti. These problems are not simply a collection of disconnected, ad hoc examples; they can be placed in a logical sequence related to the distortion of both simple and complex regulatory mechanisms. Understanding the underlying cause of problems is a first step in finding solutions, which we explore in the next chapter.

14

Rewilding

Introduction—The Extinction Crisis

In the previous chapter, we saw how the distortion of one or more of the seven subprinciples leads to the breakdown of regulation and the unraveling of ecosystems. This happens both locally and globally—in fact, the whole planet is being downgraded by the actions of humans.[1]

One of the most pressing issues of our time, apart from climate change, is the massive species-extinction crisis, also known as the sixth mass extinction in world history. The speed at which species are disappearing from the planet is at least 100 times higher than the so-called natural rate—that is, the background rate before humans had a significant impact on wildlife.[2] To put it another way, the number of species that have gone extinct in the twentieth and twenty-first centuries would have, under previous background rates, taken between 800 and 10,000 years to go extinct.[3] Extinctions have been occurring throughout geological time, but at the current rate it will take millions of years for diversity to be restored to preindustrial levels.[4] Humans have been causing the decline and extinction of species since the end of the last Ice Age, 13,000–8,000 years ago, through overhunting in combination with natural climate change. However, the current crisis is so profound that scientists are now talking about biological annihilation.[5] Around 1 million species are currently at risk of extinction.[6]

The abundance of the surviving species has also been decreasing dramatically. Humans and their livestock are now the most numerous mammals on the planet. If we measure the total weight of mammals on earth, humans represent 35 percent and livestock 60 percent, while all wild mammals combined take up only 4 percent. Chickens are three times more abundant than all wild birds combined. In most countries, the remaining large wild animals are confined to protected areas, and even there the populations are declining.[7] Between 1970 and 2012, we lost more than half of all mammals, birds, fish, amphibians, and reptiles. Large animals are the worst off, with 70 percent of species declining and 59 percent threatened with extinction. The decline in predators (75 percent worldwide) is particularly alarming, since they have such important roles in regulating ecosystems (see chapters 5 and 11). But it is not just the larger vertebrates that are disappearing. Even the smallest animals are going. In 2017, a study in Germany found an astonishing 76 percent decline in flying insects in the last 27 years. These declines were recorded in nature reserves, which indicated that protected areas could not stop the losses and showed that something more pervasive was happening.[8] Nearly all protected areas are surrounded by agricultural fields that insects are drawn to, where they succumb to pesticide use, year-round tillage of the soil, and increased use of fertilizer. Insect declines are being reported not only in Germany but in many other places in the world, and this could have far-reaching cascading effects on the food webs and ecosystem processes that provide services to humans, such as nutrient recycling and pollination, since some 80 percent of wild plants and 35 percent of the world's plant crops are pollinated by insects.

Many decades of effort in conservation have been unsuccessful in stemming the extinction crisis and decline of ecosystems. Conservation has traditionally focused on protecting nature from human intrusion and destruction by creating national parks and reserves. More recently, community-based conservation, which seeks a balance between conservation and human development, has become popular. Protected areas now cover about 19.8 million square kilometers (14.7 percent) of terrestrial land and 27.5 million square kilometers (7.6 percent) of seas and oceans.[9] This may seem like a lot, but terrestrial parks and reserves contain only about 50 percent of the total biodiversity on land; the other species are found in human-used areas outside reserves. And even inside protected areas, species are being lost as the ecosystems they live in are downgrading.

This, then, is the scope of the problem facing humanity. How do we counteract these losses and leave intact ecosystems for the next generations? We need a different and more ambitious approach to complement traditional conservation. This is where *rewilding* comes in.

Upgrading Ecosystems through Rewilding

The rinderpest story showed us how Serengeti was able to rebound from a disturbance, with profound consequences for the whole ecosystem (chapter 11). At the end of the last chapter, we saw other examples of recovering ecosystems; we called this *upgrading*. Some systems can upgrade naturally if the disturbance is removed, as was the case in Serengeti; others need help to get there. This process, which we call *trophic baseline rewilding*, is the subject of this chapter. I will use the lessons from Serengeti to illustrate this. But first, why do we need rewilding? We need healthy, regulated ecosystems to counter the extinction crisis and preserve biodiversity, but humans also benefit from services that biodiversity provides, such as clean water, climate regulation, food, and other resources. Yet there is another compelling reason why natural ecosystems containing the full trophic web are required: they act as baselines.

From 1971 to 1973, the Sahel, the savanna zone just south of the Sahara Desert, had below-average rainfall, leading climatologists to declare a drought. The area is used by pastoralists for grazing their livestock, and by 1973 there was a massive die-off of cattle and a catastrophic famine among the people, resulting in thousands of deaths. Lack of rain, it seemed, caused the deaths. However, N. H. MacLeod, a biologist at the American University in Washington, DC, noticed something odd: a satellite image of the drought-stricken country of Niger showed a patch of green within the surrounding brown, overgrazed, and denuded area.[10] This patch was a green polygon of more than 1,000 square kilometers. The polygon was a government cattle ranch, carefully managed with a rotational grazing system—in effect a modified migration. Although it had received the same low rainfall as the surrounding denuded area, the ranch was green. It differed only in the impact of the cattle and their herders. Inside the ranch, grazing was carefully controlled; outside, grazing was excessive, and soil was bare. Inside, the grass was able to grow with the available rainfall; outside, it was dead. Here was evidence that the true cause of the disaster was human-determined overgrazing, not the climate. The natural savanna, as it existed prior to 1960, was capable of withstanding the low rainfall of the early 1970s; the drought was not the cause of the famine. The lesson is that the ranch acted as a baseline, a control or reference point for the overgrazing that was happening outside. Without that baseline, we would never have understood the true causes of the famine (figure 14.1).

The green polygon highlights the fundamental need for baselines to help us understand what is happening in landscapes that have been modified by humans. We need these trophic baselines, areas that have the complete food web, as reference points against which we can compare human-use areas such as agriculture, or even partially restored areas, and managed nature reserves. Only such baselines will enable us to answer a fundamental question: Are humans causing the collapse of the ecosystems that they live in and exploit?

Some baselines are already present—parts of biomes that have not yet been radically altered. (It is understood that no area is pristine and completely undisturbed, but some minimally disturbed areas exist and can be compared against the highly exploited areas under examination.) But in many ecosystems there is no natural baseline remaining; it has all been radically altered over centuries, even millennia. In these cases we have to reconstruct the original food web in order to provide the baseline. This is what we aim for with trophic baseline rewilding.

We define **trophic baseline rewilding** as the process of rebuilding, following major human disturbance, the complete food web (trophic web) at all levels as a self-sustaining and resilient ecosystem, using biota (living organisms) that would have been present had the disturbance not occurred. This process supplements and enlarges current baselines or creates new ones if none are present. These baselines become the

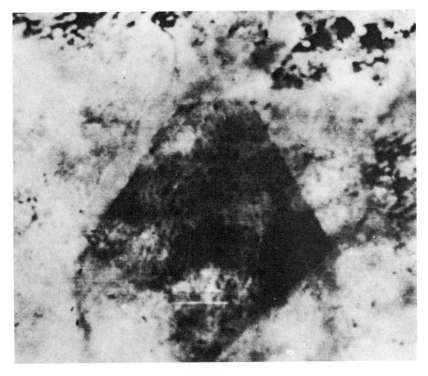

FIGURE 14.1. The "green polygon" in Niger, West Africa, seen from a NASA satellite in 1973, showing a ranch with plentiful green grass surrounded by overgrazed land. It demonstrates the value of baseline control areas by showing that it was overgrazing—not lack of rainfall—that caused the famine. (US government ERTS satellite imagery recorded by N. H. Macleod of The American University, Washington, DC.)

reference points against which human activities and all other forms of partial rewilding (discussed below), restoration, and conservation must be compared.

The definition of trophic baseline rewilding involves species that *would have been present*. This is vital because it recognizes that we are not turning back the clock to some previous arbitrary time and set of species, but are dealing with the combination of species in the undisturbed system at the present time, a combination that might have changed since the disturbance began.

We have a number of ways to identify species that would have been present. One way is to see what is present in control areas that still contain complete native communities and use these communities to build other reference areas. A second approach to finding the appropriate species is to use historical records of what was once present and would likely still be present if allowed. This approach is less certain than the first, but such records may be all that is available. A third approach is to use expert knowledge of what should be there; knowledge of indigenous peoples is often valuable in this context.

An example of the first approach is the rewilding of the Carrifran Valley, in southern Scotland, which I will revisit later.[11] This valley has been replanted with the original trees and shrubs based on similar areas still present in Norway and Sweden. This operation also used evidence from pollen analysis to determine which species were once there—the second approach. Historical records have been used in numerous cases for reintroduction of animals, such as lynx in Europe and birds on Lord Howe Island, off the coast of Australia.[12] Expert knowledge from plant ecologists identified appropriate habitats for release sites of the burrowing bettong in New South Wales,[13] and aboriginal knowledge identified sites for the release in 2019 of the mala (rufous hare wallaby) in the Tanami Desert of Australia.[14]

Trophic baseline rewilding has a second objective, in addition to recreating baselines to act as our reference points. It is to repair human-altered systems to a point where they are resilient and sustainable, using the baseline as the reference point. Such a system should be resilient to the natural disturbances with which it evolved and adaptable to a naturally changing environment. Populations in a rewilded place are regulated as they were before human-initiated disturbance caused the system to become unstable. Trophic baseline rewilding is an alternative to more human-controlled forms of conservation and restoration, which are discussed below. *It is not about what humans want, but what nature needs.*

Rewilding and Lessons from Serengeti

At this point we need to ask how the principles of ecosystem function that we documented in Serengeti can guide the process of rewilding for a baseline or for other human purposes. As a relatively intact and well-regulated system, Serengeti is a reference, a trophic baseline. The seven subprinciples underpin its regulation, and it is these principles that we can apply when repairing a downgraded ecosystem.

Restoring Bottom-Up and Top-Down Processes (Subprinciples 1 and 2)

A principal driver for the decline and extinction of species is the loss of their habitat, depriving them of food and shelter. The protection and/ or restoration of a healthy habitat is therefore fundamental to rewilding. Many ecosystems require large, little-disturbed core areas to accommodate the full food web. The Serengeti ecosystem, for example, is about 25,000 square kilometers, almost the size of Rwanda.

Reinstating bottom-up processes involves the restoration of the habitat and food sources for the animals that live or used to live there. Suitable habitat can be recreated by restoring the native vegetation, and protected core areas may be enlarged through the restoration of adjacent land and buffer zones. Missing herbivores may need to be reintroduced if they cannot recolonize spontaneously. These primary consumers then form the prey basis for secondary consumers up the food chain.

An example of rewilding through the re-creation of bottom-up processes can be found in Scotland. Although the Scottish Highlands are sometimes praised for their open spaces and seemingly wild nature, the mostly barren hills are not natural at all. They are degraded lands, the result of centuries of woodcutting, burning, and overgrazing. Organizations like Trees for Life and the Carrifran Wildwood Group of Borders Forest Trust are restoring the original Caledonian forest through rewilding.[15] They protect the land from grazing and plant the native trees, thus reinstating bottom-up processes. This is an active form of rewilding (sometimes called ecological rewilding), which means that it requires some human intervention (I'll discuss this further below).

Carrifran, the example mentioned above, is a valley in southern Scotland that has been replanted with native trees. These have attracted woodland bird species, which have been returning in a specific sequence, with insect- and seed-eating birds coming back first and raptors later. The Caledonian forest was once inhabited by large mammals such as beavers, wild boar, lynx, wolves, and brown bears, which all went extinct centuries ago. Beavers were introduced in 2009 in the Knapdale forest, and reintroduction of lynx to Scotland is being discussed.[16]

In Serengeti, we discovered how important top-down regulation by predators was in adding to the diversity and stability of the system (chapter 5). Predators, especially the large carnivores, are often missing in degraded ecosystems. One of the most famous examples of the restoration of top-down interactions through ecological rewilding is the reintroduction of wolves in Yellowstone National Park. The absence of wolves and cougars in the last century led to overbrowsing of aspen, cottonwood, and willow by elk and deer. Stream banks were particularly affected where trees and shrubs did not get a chance to regenerate. The resulting lack of vegetation led to erosion and the demise of species associated with healthy stream habitats. When wolves were reintroduced to the park in the 1990s, they brought down the numbers of elk and caused them to move around so that they spent less time near stream banks. The trees were able to recover, which reduced erosion. As the vegetation rebounded, so did the beavers and the birds that depend on riverine plants.[17] The number of beaver colonies jumped from one to nine 24 years after the wolves were reintroduced. The ponds created by the beavers also provided habitat for numerous bird and other species. Even fish benefited, as the recovered stream-bank vegetation provided shade for them.

Predators may also come back to an area naturally when the habitat becomes more suitable—that is, when hunting and other human pressures decrease and/or when prey increases. Perhaps one of the most spectacular examples of this kind of passive rewilding has been the spontaneous return of large predators in Europe. A combination of European laws protecting carnivores, abundant prey populations, land abandonment, and positive public opinion created favorable, although unintended, conditions for these animals to come back. There are now twice as many wolves (more than 11,000) in Europe than in the United States (outside Alaska), despite Europe's having twice the human density. Indeed, wolves have even been observed in densely populated countries like Belgium, the Netherlands, and Germany. There are also an estimated 17,000 brown bears and 9,000 lynx. One particular recurrence is the golden jackal, which expanded from Asia to Eastern Europe, the Balkans, and as far west as Italy and Austria. Europe is believed to support an astonishing 117,000 jackals.[18]

Protected areas are often too small to host large carnivores. That is certainly the case in Europe. If large predators had been restricted to protected areas (as in much of North America), hardly any would be left. The European practice of promoting carnivores in human landscapes overcomes the problem of small reserves. In the contiguous United States, viable populations of large carnivores persist only in large wilderness areas because of the policy among natural resource managers of extirpating carnivores in human-use areas. If carnivores were left to coexist in human landscapes, as in Europe, there would be less of a conservation problem for the top trophic level, the viability of which is a necessary condition for rewilding.[19] Successful coexistence between carnivores and humans depends on people's acceptance and knowledge of these animals, on measures to keep domestic animals safe, and on compensation to landowners whose livestock have been attacked. In areas where people and predators have been living together for centuries, problems are largely avoided by protecting sheep and other livestock with guard dogs, shepherds, the use of night corrals, and, more recently, electric fencing.

Bottom-up and top-down mechanisms of regulation are both important, and they can operate simultaneously (see the discussion of reciprocal trophic interactions in chapter 5). Many rewilding projects address both. The Ibera Nature Reserve in Northeastern Argentina is one of the largest wetlands in the world, teeming with birds and other wildlife. But despite the presence of good-quality habitat, many of the original species have disappeared. The area is a mix of protected public and private land. Around 2010, cattle were excluded from ranchland, allowing natural restoration to take place—a bottom-up intervention. Several animal species were reintroduced, including tapir, collared peccaries, pampas deer, giant anteaters, and green-winged macaws. But perhaps the most important reintroduction took place in June 2018, when two jaguar cubs were born,[20] and the ecosystem once more had a top-down predator.

In chapter 13, we described Gorongosa National Park, in Mozambique, as an example of an ecosystem being upgraded successfully. During that nation's civil war, which raged from 1977 to 1992, more than 90 percent of the animals in this park were killed. After peace returned, (now former) President Joaquim Chissano wanted to reinvigorate tourism in Mozambique's national parks. In 2004, he convinced the American businessman Gregory Carr to help restore wildlife in Gorongosa and establish nature-based tourism benefiting local communities. After the civil war, some herbivores, particularly waterbuck, increased to huge numbers, far higher than they were before the exterminations. This is clearly a distortion attributable to the absence of other herbivore and predator species. However, through active rewilding, buffalo, wildebeest, zebra, eland, and elephant were reintroduced to augment existing populations and increase genetic diversity. Wild dogs were reintroduced in 2018, and leopards came back naturally, probably from nearby hunting concessions. Lions have also increased, although hyena remain absent.[21]

Restoring Migration and Movements (Subprinciple 3)

The Great Migration is a keystone process in Serengeti, driving many parts of the ecosystem (chapter 6). While Serengeti is famous for its huge land-mammal migration, other, smaller mass migrations occur in white-eared kob in Sudan and caribou in the North American Arctic. Mass migrations once existed elsewhere too, but most have been eliminated or much reduced; two examples from history are the bison of North America and the springbok of South Africa. However, many other species, including ungulates, carnivores, birds, and even some insects, perform considerable seasonal and other movements. Lack of suitable habitat—caused by fences, roads, and human intolerance toward wildlife—often impedes movements between the areas that animals use. Restoring these movements by overcoming such barriers and providing appropriately sized habitat, as well as wildlife linkages (corridors), are fundamental tasks of rewilding.

One such project is the Yukon to Yellowstone initiative (Y2Y). Y2Y was created in 1993 to promote the protection of one of the largest rela-

tively intact mountain ecosystems in the world, 1.3 million square kilometers in size and spanning a length of 3,200 kilometers along the Rocky
Mountains between the Yukon in Northern Canada and Yellowstone in
the United States. All the species that lived there before the European
colonization of the North American continent are still present today.
With more than 450 partners, Y2Y strives to protect migrations and other
movements by improving and enlarging protected areas, creating new
ones, and allowing corridors for wildlife to move between them. In the
1990s, the project convinced the Canadian government to build 2 highway crossings and 22 underpasses to facilitate movement of wildlife across
one of the busiest highways in Western Canada, one that cuts right across
Banff National Park—an insurmountable barrier for many animals. Other
initiatives include the protection of the more than 55,000-square-kilometer
Peel watershed in the Yukon, reconnecting isolated grizzly-bear populations in Montana and Idaho, and securing wildlife corridors to support
recolonization in areas where grizzlies were extirpated. Y2Y constitutes
a vision and a movement that has inspired many other connectivity projects around the world.[22]

The Terai Arc Landscape is an 810-kilometer strip of land in northern
India and southern Nepal that connects several protected areas, including some of the most famous tiger reserves. The area is also home to the
greater one-horned rhinoceros, the Asian elephant, and numerous
other species of cats, deer, and antelope. Wildlife corridors connect
these mostly isolated, protected core areas, and forest land is being restored between them. This appears to be successful for tigers, as at least
11 individuals have been traveling through the corridors; the species is
recolonizing Nepalese protected areas from India.[23]

In Africa, migrations of wildlife were prevented by fences in Botswana,
Kruger National Park in South Africa, and elsewhere (chapter 13). Water
holes in Kruger and Hwangi in Zimbabwe kept animals in one place and
led to serious vegetation damage. In Kruger, fences between the national
park and private reserves around it and the adjacent Limpopo National
Park in Mozambique have now been taken down, and catchment dams
and artificial water holes are being decommissioned.[24] Also in southern
Africa, the Kavango Zambezi Transfrontier Conservation Area aims to

connect 36 protected areas, some of them isolated, over a region of 520,000 square kilometers shared by five countries (Angola, Botswana, Namibia, Zambia, and Zimbabwe). It supports the movements of around 200,000 elephants, the largest population in Africa.[25]

Securing corridors is often challenging, especially in densely populated areas and where there are conflicting land-use interests such as mining, agriculture, and plantations. The Mesoamerican Biological Corridor, which started as an idea to connect puma habitat in Central America (the so-called Paseo Pantera) in the 1990s, has been burdened with many challenges and conflicts. Despite endorsement by big international nongovernmental organizations (NGOs) and governmental and multilateral organizations (including the UN), as well as the expenditure of hundreds of millions of dollars, it has been slow to deliver on its promise of successfully connecting a highly fragmented landscape.[26] Building on this and other work, the Jaguar Corridor Initiative, launched by the Wildlife Conservation Society in 1999 and led by the NGO Panthera, is making some headway and provides a blueprint for connectivity between jaguar populations across Central and South America.[27]

Restoring Biodiversity (Subprinciple 4)

An important pillar of rewilding is the reintroduction, whether passive or active, of keystone species, which maintain and increase the stability and diversity of an ecosystem. I have discussed earlier how top predators were reintroduced in Ibera and Yellowstone, and self-reintroduced in Europe, but other keystones are also important, as we have seen with wildebeest in Serengeti and numerous other species (chapter 7). The following are just two examples from a list of many.

Beavers are ecosystem engineers and are as such keystone species. They create wetland habitat for many animals and plants through building dams and managing water flow. Eurasian beavers were successfully reintroduced in many European countries including the UK, the Netherlands, Belgium, Bulgaria, Austria, and Denmark,[28] while their American counterpart, a different species, has been reintroduced in several places in the American West, where they provide water habitat in a predominantly dry climate.[29]

Giant tortoises were hunted to extinction on many islands in the Pacific and Indian Ocean. They used to play an important role in the dispersion and diversity of plants. On the Galapagos and on islands near Mauritius, nonnative large tortoises were introduced as ecological replacements for the extinct species. This is an ongoing experiment, but so far there is evidence that the introduced tortoises help the dispersal of native plants and suppress invasive nonnative plants.[30]

Species are being lost when the areas in which they occur are reduced in size or isolated. Most protected areas have become or are becoming islands of nature surrounded by cultivated land. Species disappear from these "islands," but this is often a gradual process, as it takes time for the loss of habitat to ripple through the community—the phenomenon known as the extinction debt (see chapter 12).[31] It may take many decades before all future extinctions have been realized. For example, during a period of 27 to 30 years, isolated nature reserves in Ghana lost 21 to 75 percent of their large mammal species. Extinctions were initially slow but then accelerated as the effects of the disappearance of some species (like carnivores) cascaded through the system.[32] When the available habitat for wildlife drops below one-third to one-half of the landscape, species populations decline quickly because of the decreased ability of animals to move between patches. In that case, protecting and enlarging the largest remaining habitats by restoring the surrounding land is often the best conservation strategy, even if these areas are still used by humans for agriculture (albeit at a reduced rate).

Managing Disturbance (Subprinciple 5)

The interaction between vegetation and fire in combination with herbivores is often complex and can produce multiple stable states in the same ecosystem. We have seen this in Serengeti and the Maasai Mara Reserve involving wildebeest, elephants, and fire (chapter 10). Several systems in the world are adapted to fire and require it to persist. Some species need fire to disperse their seeds. For example, many forests of interior western North America experience natural fires on a regular basis. Fire suppression for commercial and safety purposes has created dense forests with a decreased biodiversity and an accumulation of highly combustible dead wood. It has also contributed, in combination with warmer winters, to epidemic outbreaks in which mountain pine beetle has killed trees in more than 18 million hectares in British Columbia alone.[33] As we saw in chapter 13, this combination led to catastrophic high-intensity fires, well beyond the low-intensity natural fires that used to burn in these forests. Restoring natural fire regimes that are frequent and of mostly low intensity is an integral part of rewilding, as is restoring other natural disturbances to which the ecosystem is adapted. Ecosystems have evolved with these disturbances—indeed, as I have indicated, disturbance may be necessary to maintain them.

In the case of fire, however, this may require large areas to preserve a mosaic of disturbed and undisturbed patches in which different species thrive in different stages of recovery. In the meantime, some management in the form of thinning, removal of high fuel loads, and controlled burning may be necessary to return the forest to a more natural regime, while taking into account human safety concerns. An example of this is Illilouette Creek basin in Yosemite National Park, where a natural fire regime has been restored. This fire-controlled forest has been shown to provide a better buffer against drought, reducing water stress on the forest and providing more water downstream than comparable fire-suppressed forests.[34]

If a disturbance is too intense, too frequent, or too long-lasting, the ecosystem may degrade and change into a different state, examples of which we saw in chapter 13. This also applies to fire, where size could be a problem. If a protected area is too small, a large, intense fire could devastate the entire area, leaving the wildlife nowhere to go. Small protected areas are more vulnerable to external influences such as invasive species; intrusion of humans and livestock; water, air, and ground pollution; and pesticides. For example, the loss of insects that has been observed in nature reserves in Germany, mentioned earlier, was likely caused by the impact of activities in the surrounding agricultural landscape. Many protected areas fall into this category; enlarging them through rewilding the surrounding landscape, by restoring native vegetation around them and creating buffer zones, can mitigate some of these disturbances.

Accommodating Continuous Change (Subprinciple 6)

Ecosystems change continuously, as we have seen in the Serengeti (chapter 9). If boundaries are rigid, natural communities cannot track shifts in climate or other environmental changes. In this way, isolated natural areas become prisons for their inhabitants, leaving them nowhere to go when conditions are no longer favorable. Accelerated climate change caused by humans makes this problem all the more urgent. How can we ensure that natural ecosystem boundaries will be flexible in the future? For this we need a large-landscape approach where biodiversity conservation is integrated within the human-use matrix.

This approach, often called *working lands conservation*, requires a radical rethinking of how we use the land to provide food, wood, minerals, and space for recreation and habitation.[35] It includes diversified farming methods, ecosystem-based forest and rangeland management, and ecological restoration of critical habitats and corridors. This should allow not only the movement of animals through wildlife corridors, as discussed in this book, but also the dispersal of plants, other organisms, and eventually whole communities.

Discontinuous Change (Subprinciple 7)

Rewilding sometimes requires moving a system from an unwanted state back to an original, preferred state. Of course, there may be situations, such as the salinization of soils in Western Australia, or the loss of soils in deforested Amazon sites, where it is no longer feasible to repair the damage. There are also situations where knowledge of the biology is not yet available for rewilding. For example, the damage to the intertidal ecosystem in southern Alaska caused by the Exxon Valdez oil spill in 1989 has still not been rectified because we have much more to learn about how actively to reintroduce marine species; there have been some advances, but there is more to learn. Nevertheless, we know from our discussion of predation (chapters 5 and 9) that predators can maintain two levels of prey numbers—two states. In some cases, one of those states is an unwanted one, and we can manipulate the populations, as part of rewilding, to allow the system to jump back to the original and desired state.

There are two opposite situations where this change to a different state occurs. The first is where nonnative predators (exotics) have invaded an ecosystem and have either reduced a vulnerable native prey to low numbers, or, in the extreme, exterminated the prey. The objective is to reduce nonnative-predator numbers to a lower level where native-prey numbers can break out and remain at a higher level even in the presence of predators. Such situations occur on islands, even huge ones like Australia and New Zealand. Red foxes and feral cats have caused the severe reduction or the extermination of many small marsupials and other types of animals in Australia.[36] What we need to do is to see if we can reduce the number of foxes so that the prey can break out to some higher number and stay there even if the predators return. There are good studies showing that black-footed rock-wallabies living on kopjes in Western Australia have rebounded after foxes were removed,[37] and a few studies exist showing some success with other species. The rufous hare-wallaby (also known as the mala) can persist in the face of fox predation if mala numbers are allowed to reach a certain threshold level. Similarly, populations of small marsupials—sandy inland mouse, spinifex hopping mouse, and numbat—are able to increase starting both from very low numbers and from a higher number, although at numbers in between these they cannot increase in the face of fox predation (this is evidence of two states, a very low density and a very high density). In New Zealand there are possible outbreaks of passerine birds (perching birds or songbirds) when introduced carnivores have been controlled. Rachelle Binny, Roger Pech, Andrea Byrom (who worked with me on rodent dynamics in Serengeti), and colleagues have shown how the removal of foreign predators can allow increases of native prey. The species most sensitive to predation and its removal are the evolutionarily oldest endemic prey. But we still do not know if any of these species can be regulated, in the face of predation, at two different levels.[38] As noted earlier (chapter 13), the Fortymile caribou herd in Alaska has been kept at low numbers (compared to historical population sizes) by wolf predation. Wolves are now being controlled and caribou numbers are increasing, but we have yet to see a clear-cut outbreak to a much higher, food-regulated level.

The second situation where we want to move from one state to another involves mainland ecosystems in which native predators were exterminated and herbivore prey populations exploded to high numbers and radically modified the vegetation. The objective is to reintroduce native predators so that they may reduce prey numbers and hold them at a lower level. There are many examples, including increases of deer species in Great Britain after wolves were exterminated in the 1700s; sika deer outbreaks in Japan after wolves disappeared in the early 1900s; and the increase of elk, moose, and deer in many places in the United States following the removal of wolves.[39] The resulting high numbers of elk in Yellowstone National Park after wolf removal in the 1920s is another case in point. Our rewilding objective is to reintroduce wolves so that they can reduce the unnaturally high level of elk and keep them at low numbers. The self-reintroduction of wolves into Banff National Park in Canada[40] and the active reintroduction of wolves into Yellowstone have significantly reduced elk numbers, and may be regulating them there. If so, rewilding has resulted in a change of state.[41]

Toward Partial Rewilding

So far, I have focused on rewilding the complete set of species and processes—*trophic baseline rewilding*. But there are many other programs that desire to reintroduce a subset of species or processes; this is called *partial rewilding*, and we should look at some of these agendas.

Ideas on how to repair damaged ecosystems have been around since at least 1990. The term *rewilding* first appeared in print in a *Newsweek* article in that year concerning radical conservation activism.[42] In 1998, Michael Soulé and Reed Noss, two eminent biologists, published a landmark article called "Rewilding and Biodiversity" in *Wild Earth* magazine.[43] They proposed to rewild large landscapes focusing on core natural areas and the connections between them. They recognized the crucial role of big predators in regulating ecosystems, and because many parks and reserves are not large enough to accommodate them, Soulé and Noss proposed connecting core areas through natural corridors. In his book *Rewilding North America: A Vision for Conservation in the 21st Century*, conservationist Dave Foreman offers rewilding as a solution to the extinction crisis: he proposes that wildland networks should reconnect, restore, and rewild the North American continent.[44] Foreman and Soulé teamed up with Doug Tompkins, the conservationist-philanthropist and cofounder of the companies The North Face and Esprit, to establish the Wildlands Project (now called Wildlands Network). This project embraced the "three C's" (core areas, corridors, and carnivores) and promoted the establishment of so-called wildways to connect North America's national parks and reserves.

Over the years, a variety of definitions of rewilding have been proposed, which has caused confusion. Some people recognized the importance of large carnivores and other keystone species and focused on their reintroduction where they had been extirpated. Others went further and proposed to introduce substitutes for now-extinct species to fill the ecological gap that was created by the extinction. All of these ideas were intended to bring a degraded system back to a more complete, self-sustaining state with restored ecological processes and species groups, while keeping human disturbances to a minimum.

Most of these proposed agendas, however, are a form of *partial rewilding* in the sense that they have a narrow or restricted set of criteria for their endpoint rather than addressing the whole trophic cascade. One agenda aims to return human-modified or degraded areas to a wilderness where there is no human interference, but without a clear definition of what wilderness is and without an obvious endpoint. Another form of partial rewilding specifies ecosystem services as the objective, but ecosystem services (meaning the benefits that humans obtain from ecosystems) are of course only a subset of all the ecological processes that are required for a stable and persistent system. There is nothing wrong with these plans or any other form of partial rewilding—some are effective for conserving species diversity—but they must be tested against a baseline with the whole set of species and processes, in case any particular agenda proves unstable in the long run.

In addition to partial rewilding, some ecologists accept or even promote the concept of novel ecosystems, which are new species assemblages often dominated by nonnative species. They argue that as long as they provide ecosystem services, we should embrace these novel ecosystems instead of spending resources on restoring them to more natural ones. But novel ecosystems are often degraded states, with an idiosyncratic endpoint that may ultimately be unstable. Again, this approach needs to be checked against a baseline, as with other agendas.

Restoration, Trophic Baseline Rewilding, and the Endpoint

Historically, ecological restoration has been primarily concerned with restoring the vegetation of degraded sites such as mine tailings or old buildings,[45] which usually did not include all the higher trophic levels of animals. More recently, however, the original purpose of restoration and the new idea of rewilding have begun converging on similar goals. The Society of Ecological Restoration now defines ecological restoration as "the process of assisting the recovery of an ecosystem that has been degraded, damaged, or destroyed."[46] This is broad and accommodates the many types of restoration practice. Objectives range from going back to a previous ecosystem with historic-species assemblages, which resembles rewilding, to establishing ecosystems with peculiar ecological features that are not entirely natural, as well as other variants, all of which amount to partial rewilding.[47] Restoration projects cover a wide range of spatial scales, including stabilizing a stream bank, revegetating a small, degraded industrial site, and restoring an entire wetland or a large degraded forest. Restoration projects often have a goal that is derived from both natural and human (i.e., cultural) values, but most of them emphasize a return to a predefined state.

Rewilding, on the other hand, is sometimes described as an open-ended process, a "trajectory to wildness in a future undefined state."[48] The shortcoming of this open-ended policy is that we have no reference baseline or endpoint to use for comparison so that we can know—or at least predict—what that food web should look like. Without this scientific control, the open-ended approach cannot be monitored for long-term sustainability, and thus it is fundamentally dangerous—it could be leading us to catastrophe without our being aware of it. We need an endpoint to guide the process of rewilding, even if the trajectory is uncertain, or socioeconomic or other constraints make it difficult to reach. We also have to be conscious that endpoints are not static but fluctuate around an equilibrium (subprinciples 1 and 5) that changes slowly over time (subprinciple 6). Long-term changes, such as human-made climate change, may cause further uncertainty about the endpoint. Short-term disturbances could lead to multiple stable states (multiple endpoints; see subprinciple 7) that can coexist in the same ecosystem, as we learned from Serengeti.[49]

Long-term monitoring in Serengeti revealed processes of which we were not aware. To be prepared for such "surprise outcomes," we need to monitor all rewilding programs, whether they address the full trophic food web, or just a part of it. Having a defined endpoint allows us to track a system's progress to the envisioned endpoint or to a different one. We can follow its trajectory, and we may intervene when it deviates because of external threats. For example, we can remove invasive species that threaten to stall the rewilding process, or protect a site from poaching.

There are two main ways to get to an identified endpoint. Perhaps the most common is *passive rewilding*, which encourages the return of species of their own accord, without much help from ecologists, to damaged ecosystems—often seen in abandoned agricultural landscapes where there is no longer any human activity or management. The return of predators in Europe, mentioned earlier, is an example of this. A second approach, which is gaining in popularity, is a more active form of rewilding called *ecological rewilding*. This is similar to the first but involves limited active management such as reintroducing lost species, especially at the beginning of the process, to assist natural recovery of the landscape. This may consist of restoring bottom-up processes, such as replanting the indigenous vegetation (as in the Carrifran Valley in

Scotland), or reinstating top-down processes by reintroducing keystone species (as with wolves in Yellowstone and several herbivores and predators in Gorongosa).

————

There have been some attempts to set endpoints for rewilding defined by conditions in prehistory. During the Late Pleistocene, for example, many large mammals roamed the North American continent, including mammoths and mastodons, giant sloths, wild horses, American lions, American cheetahs, camel-like animals, and giant tortoises. These megafauna went extinct approximately 11,000 years ago due to a combination of overhunting and a changing climate. Similar extinctions occurred in Central and South America, Australia, and Europe. Large animals had major impacts on the vegetation and biological communities by opening up dense forests, forming open glades, and changing the abundance of plant species.[50]

Some conservation biologists have proposed re-creating some of the trophic interactions that existed before these megafauna went extinct (an agenda called Pleistocene rewilding). The extinct species could not be brought back, but they could be replaced by similar species that appeared to do the same thing in the environment.[51] For example, they suggested that African cheetahs and lions, African and Asian elephants, and Bactrian camels from Central Asia could be introduced to North America to fill the ecological niches that were left vacant by the extinction of American lions, cheetahs, mastodons, and camels; these releases would benefit the environment by restoring the lost ecosystem. The drawback of this approach is that there is no guarantee that these new species would do what the lost species did or that the new artificial ecosystem would be sustainable in the present climate. Modern elephants may look like mammoths, but they evolved in the tropics and have without doubt a different ecology from that of mammoths, which evolved in the Arctic. Rare species might become endangered through a cascading effect caused by the new introductions. It is likely that over the last 11,000 years the grassland- and steppe-plant communities, where most of the extinct animals used to roam, have changed and evolved in their absence.[52]

The fact is that we don't really know how natural communities functioned in the distant past. We cannot travel back in time, map the landscape, and determine the abundance of the animals that existed then. We cannot exclude animals from plots in order to assess how the vegetation responds. It is better to stick with what we know, protect our remaining natural areas that have limited human impact, and use those as a reference for rewilding in the present time.

———

In Europe there have been proposals to return to a different historical time period—the Middle Ages. Present-day domestic animals, it was suggested, could act as substitutes for species such as the auroch (wild cow) and tarpan (wild horse) that went extinct in the 1600s. Unlike the substitutes in the Pleistocene agenda, these domestic species are at least closely related to those now extinct. One of the early projects was the Oostvaardersplassen in the Netherlands. This is an area in the north of the country that was reclaimed from the sea by the building of a huge dam. The idea was to transform the area, from scratch, into a self-sustaining ecosystem. Introduced herbivores included Heck cattle (genetically the closest domestic relative to auroch), Konik horses (related to the extinct tarpan), and red deer. These herbivores were expected to create a patchy landscape of forest and open spaces, termed woodpasture, instead of the closed-canopy forest that is now common in Central Europe.[53]

After a series of mild winters, the populations of herbivores grew considerably. When a severe winter hit, many animals starved to death, and a public outcry ensued. As a result, the authorities decided to cull a proportion of the populations before the start of the winter to spare the rest from starvation. The whole project became controversial. Proponents of the Oostvaardersplassen project claimed that starving animals were natural, while opponents argued that this was a zoo situation: the area was surrounded by a fence. The absence of predators to provide top-down regulation meant that the food web was incomplete and therefore unstable.

The problem with these approaches is that they cannot be compared to a baseline since those time periods are long gone. So we cannot test the assumption that the introduced proxies of the extinct species will have the same ecological effect as their predecessors, nor can we determine whether the approach has been successful and whether it is stable and persistent—which is the ultimate goal of rewilding.

Conditions for Rewilding

If we want to build a new ecosystem through rewilding, we must be aware of a number of conditions. First, we want present-day species, not those from a distant or arbitrary point in history because of the problems mentioned earlier. We may not know for certain, but as we explained, we can make an educated guess as to which species to return. These species must represent all trophic levels.

Second, if we want to introduce a species (or subspecies) as a replacement for an extinct species, the new one should be chosen to match the ecology of the missing one and to have the same role in the food chain. We must be aware of what might happen in the future—how the reintroduced species will perform in a changing climate. As an example, on Lord Howe Island, a remote island 600 kilometers east of Australia, nine species of birds have gone extinct since the discovery of the island by humans. One of them was the robust silvereye (*Zosterops strenua*).[54] We could replace this lost species with a genetically similar one, the silvereye from Norfolk Island (*Zosterops lateralis*), which is about 900 kilometers from Lord Howe Island. In general, species should be introduced in high enough numbers that a viable population can be established and will reach a population density that has the desired influence in the food web.[55]

Third, we need to ensure that the current ecological conditions make rewilding achievable. This will not be the case, for example, when the area has been invaded by exotic species that prevent reintroduced native species from establishing. Thus, invasive species will have to be brought under control before rewilding can take place.

Fourth, we have to make sure that rewilding is socioeconomically acceptable and feasible, or the chances of success will be slim and valuable resources will be wasted. Socioeconomic issues often arise when predators are reintroduced to areas where they had previously been exterminated. Feasibility issues are a concern in a situation like the reintroduction of rhino because rhino horn is now so valuable on the black market (at least $100,000 per kilogram, making it worth more than platinum) that poachers simply kill them using the latest technology as soon as they are introduced.

Fifth, there are occasions when we should conduct a pilot experiment in a smaller area, especially if surrogate species are being reintroduced. This will test the program assumptions and guide the main experiment. If the pilot project fails—if a surrogate species does not function as intended or a predicted trophic interaction does not occur—a bigger project will probably not work.

Our sixth and final condition is that our predictions on the outcome of the rewilding project must include measurable results. We can, for instance, predict and measure the outcome when we remove invasive species, or the changes in the biological community as a result of the introduction of top predators or the immigration of species during the recovery of habitat. In all of these cases, we must be able to predict where the new community will end up.[56]

The Take-Home Message

During the last few decades, our knowledge about ecosystems has grown spectacularly. From long-term research in the Serengeti and elsewhere, we now know how fundamental principles keep healthy communities resilient and persistent over long time periods. Throughout this book, we learned how bottom-up and top-down regulation and the diversity of species and their interactions cause ecosystems to persist and protect them against disturbances. When nature's governing principles are undermined, ecosystems unravel and may collapse. Significant downgrading had already started at the end of the Pleistocene, when humans eradicated important keystone species. However, during the last 100 years or so, as the human population has grown exponentially and the environmental footprint of each individual has increased, the decline of other species and the degradation of our ecosystems have accelerated dramatically.

We are now witnessing a downgrading of ecosystems all over the world because of overexploitation of resources, loss of habitat, overhunting, human-made climate change, and other human-caused disturbances. This downgrading threatens the life-support systems of hundreds of thousands of species and ultimately of human beings, since we all depend on the services that ecosystems provide.

We need to reverse course. First, we need baselines of healthy ecosystems against which we can compare human-modified systems. Only with controls can we assess what does and does not lead to persistent, healthy, and resilient systems, and whether we are heading for collapse or not.

Second, we must make sure that a large area of the planet is and will continue to be able to support the multitude of species and interactions that make ecosystems diverse, productive, self-sustaining, and resilient. With impending drastic changes to our climate, this becomes all the more important. As we have seen, ecosystems that are more resilient have a better chance to weather these changes. It is also shortsighted to address climate change without considering ecosystems, because this is where the impact will be felt, and this is where it matters.

Protecting our last remaining wildernesses or fragments of nature is not enough to provide the baselines that we need and to prevent further loss of species and downgrading of our planet. We need to rewild—to rebuild trophic cascades through protection, restoration of habitat, the creation of corridors, and the reintroduction of species, keystone species above all. Because life depends on it, investing in the recovery of ecosystems should be at the top of any agenda for humanity in the twenty-first century and beyond. It is our moral obligation to future generations.

Mammals and Trees Mentioned in the Text

Chiroptera

Straw-colored fruit bat   *Eidolon helvum*

Afrotheria

Savanna elephant   *Loxodonta africana*
Forest elephant   *Loxodonta cyclotis*
Asian elephant   *Elephas maximus*
Bush hyrax   *Heterohyrax brucei*
Rock hyrax   *Procavia capensis*
Tree hyrax   *Dendrohyrax arboreus*
Aardvark   *Orycteropus afer*

Rodentia

Springhare   *Pedetes surdaster*
Brandt's vole   *Microtus brandti*
Pacific rat (kiore)   *Rattus exulans*

Lagomorpha

European rabbit   *Oryctolagus cuniculus*
Cape hare   *Lepus capensis*

Pholidota

Ground pangolin   *Smutsia temminckii*

Pilosa

Giant anteater   *Myrmecophaga tridactyla*

Primates

Guereza black-and-white colobus   *Colobus guereza*
Olive baboon   *Papio anubis*

Carnivora

African wild dog   *Lycaon pictus*
Black-backed jackal   *Lupulella (Canis) mesomelas*
Side-striped jackal   *Lupulella (Canis) adusta*
African golden wolf   *Canis lupaster*
Asian golden jackal   *Canis aureus*
Coyote   *Canis latrans*
Wolf   *Canis lupus*
Island fox   *Urocyon littoralis*
Bat-eared fox   *Otocyon megalotis*
Ratel, honey badger   *Mellivora capensis*
Zorilla   *Ictonyx striata*
Striped weasel   *Poecilogale albinucha*
Pacific sea otter   *Enhydra lutris*
Island skunk   *Spilogale gracilis amphiala*
Common genet   *Genetta genetta*

Civet, African	*Civettictis civetta*
Lion	*Panthera leo*
Leopard	*Panthera pardus*
Jaguar	*Panthera onca*
European lynx	*Lynx lynx*
Cheetah	*Acinonyx jubatus*
Spotted hyena	*Crocuta crocuta*
Giant panda	*Ailuropoda melanoleuca*
Brown bear	*Ursus arctos*
Steller's sea lion	*Eumetopias jubatus*

Perissodactyla

Grant's zebra	*Equus quagga boehmi*
Black rhinoceros	*Diceros bicornis*
White rhinoceros	*Ceratotherium simum*
Greater one-horned rhinoceros	*Rhinoceros unicornis*
South American tapir	*Tapirus terrestris*
Tarpan	*Equus ferus ferus*

Artiodactyla

Collared peccary	*Pecari tajacu*
Giant forest hog	*Hylochoerus meinertzhageni*
Warthog	*Phacochoerus africanus*
Hippopotamus	*Hippopotamus amphibius*
Maasai giraffe	*Giraffa tippelskirchi*
African buffalo	*Syncerus caffer*
Auroch	*Bos primigenius*
North American bison	*Bison bison*
Lesser kudu	*Ammelaphus (Tragelaphus) imberbis*
Eland	*Tragelaphus (Taurotragus) oryx*
Bushbuck	*Tragelaphus scriptus*

Sitatunga	*Tragelaphus spekii*
Greater kudu	*Strepsiceros zambesiensis*
Grey duiker	*Sylvicapra grimmia*
Roan antelope	*Hippotragus equinus*
Oryx (Fringe-eared)	*Oryx beisa*
Steinbuck	*Raphicerus campestris*
Dikdik (Kirk's)	*Madoqua kirkii*
Oribi	*Ourebia ourebi*
Klipspringer	*Oreotragus oreotragus*
Thomson's gazelle	*Eudorcas thomsonii*
Grant's gazelle	*Nanger granti*
Defassa waterbuck	*Kobus ellipsiprymnus defassa*
Bohor reedbuck	*Redunca redunca*
Mountain reedbuck	*Redunca fulvorufula*
White-eared kob	*Kobus kob*
Lechwe (Red)	*Kobus leche*
Impala	*Aepyceros melampus*
Topi	*Damaliscus lunatus jimela*
Kongoni (Coke's hartebeest)	*Alcelaphus buselaphus cokii*
Wildebeest	*Connochaetes taurinus*
Saiga antelope	*Saiga tatarica*
Caribou	*Rangifer tarandus*
Moose	*Alces alces*
Elk	*Cervus canadensis*
Red deer	*Cervus elaphus*
Roe deer	*Capreolus capreolus*
White-tailed deer	*Odocoileus virginianus*
Pampas deer	*Ozotoceros bezoarticus*

Cetacea

Killer whale (orca)	*Orcinus orca*
North Atlantic right whale	*Eubalaena glacialis*

Marsupialia

Burrowing bettong	*Bettongia lesurur*
Mala (rufous hare-wallaby)	*Lagorchestes hirsutus*
Black-footed rock-wallaby	*Petrogale lateralis*
Numbat	*Myrmecobius fasciatus*
Sandy inland mouse,	*Pseudomys hermannsburgensis*
Spinifex hopping mouse	*Notomys alexis*

Trees and Shrubs

Forest

Giant diospyros	*Diospyros abyssinica*
Forest ironwood	*Drypetes gerrardii*
Elaeodendron	*Cassine (Elaeodendron) buchananii*
Cedar	*Afrocarpus (Podocarpus) usambarensis*
Tamarind tree	*Tamarindus indica*
Grey-bark saucer-berry	*Cordia goetzei*
Euclea	*Euclea divinorum*
Furry-fruited teclea	*Teclea trichocarpa*
Wild olive	*Olea europaea africana*

Broad-leaved woodland

Large-leaved terminalia	*Terminalia mollis*
Velvet bushwillow	*Combretum molle*
Marula	*Sclerocarya birrea*
Lance-pod	*Lonchocarpus eriocalyx*
Tar berry	*Ozoroa (Heeria) reticulate*
Rhus	*Rhus natalensis*

Savanna

Umbrella acacia	*Vachellia (Acacia) tortilis*
Stink-bark acacia	*Vachellia (Acacia) robusta*
Yellow fever acacia	*Vachellia (Acacia) xanthophloea*
Whistling thorn acacia	*Vachellia (Acacia) drepanolobium*
Silver whistling thorn acacia	*Vachellia (Acacia) seyal*
White-thorn acacia	*Vachellia (Acacia) hockii*
White river thorn acacia	*Senegalia (Acacia) polyacantha*
Gum or three-thorned acacia	*Senegalia (Acacia) senegal*
African date	*Balanites aegyptiaca*
Gardenia	*Gardenia ternifolia jovis-tonantis*
Fig	*Ficus* species
Sausage tree	*Kigelia africana*
Candelabra tree	*Euphorbia candelabrum*
Caterpillar bush	*Ormocarpum trichocarpum*
Sicklebush	*Dichrostachys cinerea*
African myrrh	*Commiphora africana*
Sandpaper bush	*Cordia ovalis*
Grewia	*Grewia trichocarpa*

NOTES

Chapter 1: Why Serengeti?

1. In fact, this idea was originally proposed in 1965 by Russell Train (a Republican environmentalist who became head of the Environmental Protection Agency) at a White House conference in Washington, DC; he called for a "World Heritage Trust" that would protect "the world's superb natural and scenic areas and historic sites for the present and the future of the entire world citizenry." In 1968, the International Union for the Conservation of Nature (IUCN) developed similar proposals for its members. See https://whc.unesco.org/en/convention. Director of the World Heritage Sites Mechtild Rössler and Professor Christina Cameron of the University of Montréal have written on their history. See Cameron, Christina, and Mechtild Rössler. (2016) *Many Voices, One Vision: The Early Years of the World Heritage Convention.* Oxfordshire: Routledge. https://www.routledge.com/Many-Voices-One-Vision-The-Early-Years-of-the -World-Heritage-Convention/Cameron-Rossler/p/book/9781138248083.

2. Peters, C. R., R. J. Blumenschine, R. L. Hay, et al. (2008) "Paleoecology of the Serengeti-Mara Ecosystem." In A. R. E. Sinclair, C. Packer, S. A. R. Mduma, and J. M. Fryxell (eds.), *Serengeti III: Human Impacts on Ecosystem Dynamics.* Chicago: University of Chicago Press: 47–94.

3. Grzimek, B., and M. Grzimek. (1960) *Serengeti Shall Not Die.* London: Hamish Hamilton, Ltd. (documentary film).

4. The sociopolitical events that determined the history of conservation in Serengeti are documented in Sinclair, A. R. E. (2012) *Serengeti Story.* Oxford: Oxford University Press.

5. Geography and geology are documented in Metzger, K. L., A. R. E. Sinclair, A. Macfarlane, M. B. Coughenour, and J. Ding. (2015) "Scales of change in the Greater Serengeti ecosystem." In A. R. E. Sinclair, K. L. Metzger, J. M. Fryxell, and S. A. R. Mduma (eds.), *Serengeti IV: Sustaining Biodiversity in a Coupled Human-Natural System.* Chicago: University of Chicago Press: 33–72.

6. For information on the riverine forests, see Sharam, G., A. R. E. Sinclair, and R. Turkington. (2006) "Establishment of Broad-Leaved Thickets in Serengeti, Tanzania: The Influence of Fire, Browsers, Grass Competition, and Elephants." *Biotropica* 38: 599–605; Sharam, G., A. R. E. Sinclair, R. Turkington, and A. L. Jacob. (2009) "The Savanna Tree *Acacia polyacantha* Facilitates the Establishment of Riparian Forests in Serengeti National Park, Tanzania." *Journal of Tropical Ecology* 25: 31–40; Sharam, G., A. R. E. Sinclair, and R. Turkington. (2009) "Serengeti Birds Maintain Forests by Inhibiting Seed Predators." *Science* 325: 51.

7. On what drives the migration, see Fryxell, J. M. (1995) "Aggregation and Migration by Grazing Ungulates in Relation to Resources and Predators." In A. R. E. Sinclair and P. Arcese (eds.), *Serengeti II: Dynamics, Management, and Conservation of an Ecosystem.* Chicago: University

of Chicago Press: 257–273; Murray, M. G. (1995) "Specific Nutrient Requirements and Migration of Wildebeest." In A. R. E. Sinclair and P. Arcese (eds.), *Serengeti II: Dynamics, Management, and Conservation of an Ecosystem*. Chicago: University of Chicago Press: 231–256; Fryxell, J. M., J. F. Wilmshurst, A. R. E. Sinclair, D. T. Haydon, R. D. Holt, and P. A. Abrams. (2005) "Landscape Scale, Heterogeneity, and the Viability of Serengeti Grazers." *Ecology Letters* 8: 328–335; Fryxell, J. M., J. F. Wilmshurst, and A. R. E. Sinclair. (2004) "Predictive Models of Movement by Serengeti Grazers." *Ecology* 85: 2429–35; Holdo, R. M., R. D. Holt, and J. M. Fryxell. (2009) "Opposing Rainfall and Nutrient Gradients Best Explain the Wildebeest Migration in the Serengeti." *American Naturalist* 173: 431–45.

8. For facilitation by wildebeest on gazelle, see McNaughton, S. J. (1976) "Serengeti Migratory Wildebeest: Facilitation of Energy Flow by Grazing." *Science* 191: 92–94.

9. For impala behavior, see Jarman, P. J., and M. V. Jarman. (1973) "Social Behaviour, Population Structure, and Reproductive Potential in Impala." *East African Wildlife Journal* 11: 329–338.

10. For papers on oribi in Serengeti, see Arcese, P., G. Jongejan, and A. R. E. Sinclair. (1995) "Behavioural Flexibility in Small African Antelope: The Size and Composition of Oribi Groups." *Ethology* 99: 1–23; Mduma, S. A. R. (1995) "Distribution and Abundance of Oribi, a Small Antelope." In A. R. E. Sinclair and P. Arcese (eds.), *Serengeti II: Dynamics, Management, and Conservation of an Ecosystem*. Chicago: University of Chicago Press: 220–230.

11. For buffalo habitat choice and distribution, see Sinclair, A. R. E. (1977) *The African Buffalo. A Study of Resource Limitation of Populations*. Chicago: University of Chicago Press: 56–60, 202.

12. Strauss, M. K. L., M. Kilewo, D. Rentsch, and C. Packer. (2015) "Food supply and poaching limit giraffe abundance in the Serengeti." *Population Ecology* doi:10.1007/s10144-015-0499-9.

13. Publications on lions in Serengeti are numerous. See especially Schaller, G. B. (1972) *The Serengeti Lion*. Chicago: University of Chicago Press; Bertram, B.C.R. (1973) "Lion Population Regulation." *East African Wildlife Journal* 11: 215–225; Hanby, J. P., J. D. Bygott, and C. Packer. (1995) "Ecology, Demography, and Behaviour of Lions in Two Contrasting Habitats: Ngorongoro Crater and Serengeti Plains." In A. R. E. Sinclair and P. Arcese (eds.), *Serengeti II: Dynamics, Management, and Conservation of an Ecosystem*. Chicago: University of Chicago Press: 315–331; Hopcraft, J. G. C., A. R. E. Sinclair, and C. Packer. (2005) "Prey Accessibility Outweighs Prey Abundance for the Location of Hunts in Serengeti Lions." *Journal of Animal Ecology* 74: 559–566; Packer, C., R. Hilborn, A. Mosser, et al. (2005) "Ecological Change, Group Territoriality, and Population Dynamics in Serengeti Lions." *Science* 307: 390–393.

14. Important publications on hyenas are Kruuk, H. (1972) *The Spotted Hyena*. Chicago: University of Chicago Press; Frank, L. G., K. E. Holekamp, and L. Smale. (1995) "Dominance, Demography, and Reproductive Success of Female Spotted Hyena." In A. R. E. Sinclair and P. Arcese (eds.), *Serengeti II: Dynamics, Management, and Conservation of an Ecosystem*. Chicago: University of Chicago Press: 364–384; Hofer, H., and M. East. (1995) "Population Dynamics, Population Size, and the Commuting System of Serengeti Spotted Hyena." In A. R. E. Sinclair and P. Arcese (eds.), *Serengeti II: Dynamics, Management, and Conservation of an Ecosystem*. Chicago: University of Chicago Press: 332–363.

15. Afrotheria may have evolved from the ancestral Afrothere group entirely within Africa possibly after the end of the Cretaceous mass extinction. It was only recognized as a group in the late 1990s. See Stanhope, M. J., V. G. Waddell, O. Madsen, et al. (1998) "Molecular evidence

for multiple origins of Insectivora and for a new order of endemic African insectivore mammals." *Proceedings of the National Academy of Sciences, USA* 95: 9967–9972.

16. Bird, M. I., and J. A. Cali. (1998) "A million-year record of fire in sub-Saharan Africa." *Nature* 394: 767–769.

17. Metzger, K. L., A. R. E. Sinclair, A. Macfarlane, M. B. Coughenour, and J. Ding. (2015) "Scales of Change in the Greater Serengeti Ecosystem." In A. R. E. Sinclair, K. L. Metzger, J. M. Fryxell, and S. A. R. Mduma (eds.), *Serengeti IV: Sustaining Biodiversity in a Coupled Human-Natural System.* Chicago: University of Chicago Press: 32–72.

18. Sinclair, A. R. E., K. L. Metzger, J. M. Fryxell, et al. (2013) "Asynchronous food web pathways buffer the response of Serengeti predators to El Niño Southern Oscillation." *Ecology* 94: 1123–1130.

19. Mduma, S., R. Hilborn, and A. R. E. Sinclair. (1998) "Limits to Exploitation of Serengeti Wildebeest and Implications for Its Management." In D. M. Newbery, N. Brown, and H. H. T. Prins (eds.), *Dynamics of Tropical Communities.* British Ecological Society Symposium 37: 243–265. Oxford: Blackwell Science.

20. Sinclair, A.R.E. (2012) *Serengeti Story.* Oxford: Oxford University Press.

Chapter 2: The Discovery of Rinderpest

1. My job was to record the bird migrations from Asia as assistant to Professor A. J. Cain. He had projects elsewhere in Africa and left me to it for three months. The answer to the question is that migrants could only fit into the Serengeti ecosystem after the short rains in November brought large numbers of migrating insects, which the migrating birds could feed upon. Before rains, the birds could not stay and passed on to find areas where rains had occurred. See Sinclair, A.R.E. (1978) "Factors affecting the food supply and breeding season of resident birds and movements of Palaearctic migrants in a tropical African savannah." *Ibis* 120: 480–497.

2. Pankhurst, R. (1966) "The Great Ethiopian Famine of 1888–1892: A New Assessment. Part 2." *Journal of the History of Medicine* July: 271–94; Waller, R. D. (1988) "Emutai: Crisis and Response in Maasailand 1883–1902." In D. Johnson and D. Anderson (eds.), *The Ecology of Survival: Case Studies from Northeast African History.* Boulder, CO: Westview Press: 73–114; Spinage, C. A. (2003) *Cattle Plague: A History.* Dordrecht: Kluwer Academic/Plenum Press.

3. From Kjekshus, H. (1977) *Ecology Control and Economic Development in East Africa.* London: Heinemann Educational Books, as reported in Coulson, A. (1982) *Tanzania: A Political Economy.* Oxford: Clarendon Press: 25.

4. E. B. Worthington, working for the Food and Agriculture Organization in Rome, understood that rinderpest was the single most serious impediment to animal husbandry and food production in Africa. He assumed, like everyone else at the time, that rinderpest came from wildlife. So he recommended the start of wildlife research in 1951 (letter in the Beaty Biodiversity Research Centre archives). The first person appointed by the British Colonial Office in this program for the Serengeti was Allan Brooks, a Canadian from British Columbia. He began work on Thomson's gazelle in 1952. This was the first wildlife research in East Africa. Worthington and his family came to help Brooks carry out the very first ground-transect census of wildlife on the plains in September 1952 and February 1953.

5. Baumann, O. (1894, repr. 1968) *Durch Massailand zur Nilquelle*. New York: Johnson Reprint Corp.

6. Smith, G. E. (1907) "From the Victoria Nyanza to Kilimanjaro." *Geographical Journal* 29: 249–69.

7. Speke, J. H. (1863) *Journal of the Discovery of the Sources of the Nile*. London and Edinburgh: Harper and Bros.

8. Baumann explicitly used the word *Seuche*, meaning epidemic (p. 36). See note 5.

9. Quoted in Lydekker, R. (1908) *The Game Animals of Africa*. London: Rowland Ward; see also Millais, J. G. (1918; repr. 2006) *Life of Frederick Courtenay Selous*. Zanzibar: Gallery Publications; Hinde, S. L., and H. Hinde. (1901) *The Last of the Masai*. London: Heinemann.

10. Roosevelt, T. (1910) *African Game Trails*. London: John Murray.

11. Percival, A. B. (1924) *A Game Ranger's Note Book*. London: Nisbet and Co.

12. Moore, A. (1938) *Serengeti*. London: Country Life.

13. Ford, J. (1971) *The Role of the Trypanosomiases in African Ecology*. Oxford: Clarendon Press.

14. The burning policy is reported in Huxley, J. (1936) *Africa View*. London: Chatto and Windus.

15. Walter Plowright began work on his PhD in 1960 and was the leading virologist in understanding rinderpest. Plowright, W. (1963) "The role of game animals in the epizootiology of rinderpest and malignant catarrhal fever in East Africa." *Bulletin of Epizootiological Disease in Africa* 11: 149–162; Plowright, W. (1982) "The effects of rinderpest and rinderpest control on wildlife in Africa." *Symposium of the Zoological Society, London* 50: 1–28.

16. The data for rinderpest antibodies can be found in Plowright, W., and B. McCullough. (1967) "Investigations on the incidence of rinderpest virus infection of game animals of N. Tanzania and S. Kenya." *Journal of Hygiene, Cambridge* 65: 343–358; Taylor, W. P., and Watson, R. M. (1967) "Studies on the epizootiology of rinderpest in blue wildebeest and other game species of northern Tanzania and southern Kenya, 1965–67." *Journal of Hygiene, Cambridge* 65: 537–45; Atang, P. G., and W. Plowright. (1969) "Extension of the JP-15 rinderpest control campaign to Eastern Africa: The epizootiological background." *Bulletin of Epizootiological Disease in Africa* 17: 161–170; Rossiter, P. B., W. P. Taylor, B. Bwangamoi, A. R. H. Ngereza, et al. (1987) "Continuing presence of rinderpest virus as a threat in East Africa, 1983–1985." *Veterinary Record* 120: 59–62.

17. The rinderpest antibody frequencies relative to the year of birth are given in Sinclair, A. R. E. (1977) *The African Buffalo*. Chicago: University of Chicago Press.

18. Andy Dobson has diverse interests, from population modeling to epidemiology, biodiversity, conservation, and even the biopolitics of the ivory trade. See Dobson, A. (1995) "The ecology and epidemiology of rinderpest virus in Serengeti and Ngorongoro Conservation Area." In A. R. E. Sinclair and P. Arcese (eds.), *Serengeti II: Dynamics, Management, and Conservation of an Ecosystem*. Chicago: University of Chicago Press: 485–505.

Chapter 3: Finding Regulation

1. Early buffalo total counts were conducted by up to 6 planes with park wardens from many parks coming to help. Early wildebeest transect counts were conducted by Mike Norton-Griffiths and me in 1971–78 and then Markus Borner of FZS from 1980 to the early 2000s. Grant Hopcraft took over from 2003 onward.

2. Buffalo census techniques are in Sinclair, A. R. E. (1973) "Population increases of buffalo and wildebeest in the Serengeti." *East African Wildlife Journal* 11: 93–107.

3. For the first surveys, see Grzimek, M., and B. Grzimek. (1960) "A Study of the Game of the Serengeti Plains." *Zeitschrift für Säugetierkunde* 25: 1–61; Stewart, D. R. M. (1962) "Census of Wildlife on the Serengeti, Mara and Loita Plains." *East African Agricultural and Forestry Journal* 28: 58–60.

4. The wildebeest census method is given in Norton-Griffiths, M. (1973) "Counting the Serengeti migratory wildebeest using two-stage sampling." *East African Wildlife Journal* 11: 135–149. Computer-recognition counting is reported in Torney, C. J., D. J. Lloyd-Jones, M. Chevallier, et al. (2019) "A comparison of deep learning and citizen science techniques for counting wildlife in aerial survey images." *Methods in Ecology & Evolution* 2019: 1–9. doi: 10.1111/2041–210X.13165.

5. Malthus, T. R. (1798) *An Essay on the Principle of Population.* London: Johnson.

6. Early papers proposed that weather, or a variety of factors that depended on weather, limited animal population numbers. Andrewartha and Birch developed the theory of environmental limiting factors in Australia, where both weather and populations fluctuate markedly. Davidson, J., and H. G. Andrewartha. (1948) "Annual trends in a natural population of *Thrips imaginis* (Thysanoptera)." *Journal of Animal Ecology* 17: 193–199; Andrewartha, H. G., and L. C. Birch. (1954) *The Distribution and Abundance of Animals.* Chicago: University of Chicago Press.

7. Random events: If you use a computer to generate a series of random numbers between zero and, say, 600, and stop the series every time it reaches zero or 600 (simulating extinction), then the length of the run (which we can think of as years) will vary on average anywhere from 100 to about 200 years. If we allow an upper boundary of 600 (meaning that the computer must choose a number below 600 for the next time period) and only accept extinction when it reaches zero, then the longevity of the runs is much greater—several hundred years. If you put boundaries on both sides so that numbers cannot go below 1, then the population will extend forever. By establishing boundaries, we are in fact asserting regulation in a very extreme form— none in between the boundaries, and then extreme negative feedback at the boundary. It turns out that there were boundaries to the thrips populations studied by Davidson and Andrewartha (see note 6, above), so the populations were regulated. However, in nature there are other processes that take over when populations reach very low numbers: I deal with these in a later chapter.

8. A. J. Nicholson revolutionized thinking about population limitation when he introduced the concept of regulation through density-dependent factors. Factors that could act in this way include predation or competition between animals for food or space. However, there was little evidence in nature on these processes, and those who championed the weather ideas discounted the regulation theory. By 1958, there was much debate but little evidence one way or the other. See Nicholson, A. J. (1933) "The Balance of Animal Populations." *Journal of Animal Ecology* 2: 132–178. I detail the history of the regulation theory in Sinclair, A. R. E. (1989) "Population Regulation in Animals." In J. M. Cherrett (ed.), *Ecological Concepts.* Oxford: Blackwell Scientific: 197–241.

9. The theory of regulation is explained in detail in chapter 5 of Fryxell, J., A. R. E. Sinclair, and G. Caughley. (2014) *Wildlife Ecology, Conservation, and Management.* Oxford: Wiley Blackwell, 3rd ed.

10. Early studies of bird populations were led by David Lack at the Edward Grey Institute, Oxford University. See Lack, D. (1954) *The Natural Regulation of Animal Numbers.* Oxford: Clarendon Press; Lack, D. (1955) "Mortality factors affecting adult numbers." In J. B. Cragg and N. W. Pirie (eds.), *The Numbers of Man and Animals.* The Institute of Biology. London: Oliver & Boyd: 47–56; Lack, D. (1966) *Population Studies of Birds.* Oxford: Clarendon Press.

11. A couple of papers by the main proponents of regulation by social behavior include Krebs, C. J., and J. Myers. (1974) "Population cycles in small mammals." *Advances in Ecological Research* 8: 267–399; Wynne-Edwards, V. C. (1965) "Social organization as a population regulator." In P. E. Ellis (ed.), *Social Organization of Animal Communities. Symposium of the Zoological Society of London* 14: 173–178.

12. The theory of regulation and evidence for density dependence in Serengeti buffalo is given in Sinclair, A. R. E. (1974) "The natural regulation of buffalo populations in East Africa. IV. The food supply as a regulating factor, and competition." *East African Wildlife Journal* 12: 291–311; and synthesized in Sinclair, A. R. E. (1977) *The African Buffalo.* Chicago: University of Chicago Press.

13. Early census results are given in Sinclair, A. R. E. (1977) *The African Buffalo.* Chicago: University of Chicago Press; Sinclair, A. R. E. (1979) "The eruption of the ruminants." In A. R. E. Sinclair and M. Norton-Griffiths (eds.), *Serengeti: Dynamics of an Ecosystem.* Chicago: University of Chicago Press: 82–103; Sinclair, A. R. E., J. G. C. Hopcraft, H. Olff, S. A. R. Mduma, K. A. Galvin, and G. J. Sharam. (2008) "Historical and future changes to the Serengeti ecosystem." In A. R. E. Sinclair, C. Packer, S. A. R. Mduma, and J. M. Fryxell (eds.), *Serengeti III: Human Impacts on Ecosystem Dynamics.* Chicago: University of Chicago Press: 7–46; Hopcraft, J. G. C., R. M. Holdo, E. Mwangomo, et al. (2015) "Why are wildebeest the most abundant herbivore in the Serengeti ecosystem?" In A. R. E. Sinclair, K. L. Metzger, J. M. Fryxell, and S. A. R. Mduma (eds.), *Serengeti IV: Sustaining Biodiversity in a Coupled Human-Natural System.* Chicago: University of Chicago Press: 125–174.

14. Markus Borner led the Frankfurst Zoological Society program in Serengeti from 1984 to 2012, when he retired. He was instrumental in funding management during the bleak years of 1977–86, when the border with Kenya was closed. He helped with most of the censuses during this time.

15. Simon Mduma first joined the team in 1988 for a master's degree on the oribi. He went on to analyze the regulation of the wildebeest population, a study that he completed in 1996, and then became the principal investigator for our Serengeti Biodiversity Program. In 2005 he was promoted to director-general of the Tanzania Wildlife Research Institute, which oversees all wildlife research in Tanzania. Since retiring in 2020, he has continued to supervise our Serengeti Biodiversity Program.

16. Evidence for the regulation of wildebeest is given in Sinclair, A. R. E. (1977) *The African Buffalo.* Chicago: University of Chicago Press; Sinclair, A. R. E., H. Dublin, and M. Borner. (1985) "Population regulation of Serengeti wildebeest: a test of the food hypothesis." *Oecologia* 65: 266–268; Mduma, S. A. R., A. R. E. Sinclair, and R. Hilborn. (1999) "Food regulates the Serengeti wildebeest population: a 40-year record." *Journal of Animal Ecology* 68: 1101–1122.

17. Ray Hilborn, while a research associate at the University of British Columbia, collaborated with me on modeling wildebeest regulation. He later went to the School of Fisheries, University of Washington, and collaborated with me on wildebeest dynamics from the 1970s to the 2010s.

18. See Hilborn, R., and A. R. E. Sinclair. (1979) "A simulation of wildebeest and other un-gulates and their predators in the Serengeti." In A. R. E. Sinclair and M. Norton-Griffiths (eds.), *Serengeti: Dynamics of an Ecosystem*. Chicago: University of Chicago Press: 287–309.

19. The collapse of buffalo numbers due to poaching and their subsequent return are given in Dublin, H. T., A. R. E. Sinclair, S. Boutin, E. Anderson, M. Jago, and P. Arcese. (1990) "Does Competition Regulate Ungulate Populations? Further Evidence from Serengeti, Tanzania." *Oecologia* 82: 238–288; Metzger, K. L., A. R. E. Sinclair, R. Hilborn, J. G. C. Hopcraft, and S. A. R. Mduma. (2010) "Evaluating the protection of wildlife in parks: the case of African buffalo in Serengeti." *Biodiversity & Conservation* 19: 3431–3444.

20. See note 7.

21. See Sinclair, A. R. E. (1977) *The African Buffalo*. Chicago: University of Chicago Press.

22. Over 100 studies showing regulation are given in Fowler, C. W. (1987) "A review of density dependence in populations of large mammals." In H. H. Genoways (ed.), *Current Mammalogy*. New York: Plenum Press: 401–441. Some specific examples are white-eared kob: Fryxell, J. M. (1987) "Food Limitation and the Demography of a Migratory Antelope, the White-Eared Kob." *Oecologia*, 72: 83–91; greater kudu: Owen-Smith, N. (1990) "Demography of a Large Herbivore, the Greater Kudu *Tragelaphus strepsiceros*, in Relation to Rainfall." *Journal of Animal Ecology* 59: 893–913; bighorn sheep: Festa-Bianchet, M., J. M. Gaillard, and J. T. Jorgenson (1998) "Mass and Density-Dependent Reproductive Success and Reproductive Costs in a Capital Breeder." *American Naturalist* 152: 367–379; Soay sheep: Coulson, T., E. A. Catchpole, S. D. Albon, et al. (2001) "Age, Sex, Density, Winter Weather, and Population Crashes in Soay Sheep." *Science* 292: 1528–1531; elephant in Kruger National Park: Sinclair, A. R. E., and K. Metzger. (2009) "Advances in wildlife ecology and the influence of Graeme Caughley." *Wildlife Research* 36: 8–15. See also note 6.

Chapter 4: The Discovery of Food Regulation

1. Cheatum, E. L. (1949) "Bone marrow as an index of malnutrition in deer." *New York State Conservationist* 3: 19–22.

2. Kidney and bone-marrow fat are reported in Sinclair, A. R. E., and P. Duncan. (1972) "Indices of condition in tropical ruminants." *East African Wildlife Journal* 10: 143–149; Sinclair, A. R. E. (1977) *The African Buffalo. A Study of Resource Limitation of Populations*. Chicago: University of Chicago Press; and synthesized in chapters 2 and 5 of Fryxell, J., A. R. E. Sinclair, and G. Caughley. (2014) *Wildlife Ecology, Conservation, and Management*. Oxford: Wiley Blackwell, 3rd ed. See also Hanks, J. (1981) "Characterization of population condition." In C. W. Fowler, and T. D. Smith (eds.), *Dynamics of Large Mammal Populations*. New York: Wiley & Sons: 47–73.

3. Seasonality of fat deposition in wildebeest is documented in Sinclair, A. R. E. (1977) *The African Buffalo. A Study of Resource Limitation of Populations*. Chicago: University of Chicago Press.

4. See note 3.

5. Sinclair, A. R. E., H. Dublin, and M. Borner. (1985) "Population regulation of Serengeti wildebeest: a test of the food hypothesis." *Oecologia* 65: 266–268. Regulation is synthesized in chapter 2 of Fryxell, J., A. R. E. Sinclair, and G. Caughley. (2014) *Wildlife Ecology, Conservation, and Management*. Oxford: Wiley Blackwell, 3rd ed.

6. Mduma, S. A. R., A. R. E. Sinclair, and R. Hilborn. (1999) "Food regulates the Serengeti wildebeest population: a 40-year record." *Journal of Animal Ecology* 68: 1101–1122.

7. Craig Packer's long-term studies are summarized in Packer, C., R. Hilborn, A. Mosser, et al. (2005) "Ecological Change, Group Territoriality and Population Dynamics in Serengeti Lions." *Science* 307: 390–393.

8. Where lions choose to hunt is described in Hopcraft, J. G. C., A. R. E. Sinclair, and C. Packer. (2005) "Prey accessibility outweighs prey abundance for the location of hunts in Serengeti lions." *Journal of Animal Ecology* 74: 559–566.

9. Sinclair, A. R. E., and P. Arcese. (1995) "Population consequences of predation sensitive foraging: the Serengeti wildebeest." *Ecology* 76: 882–891.

10. Joseph Osgood of Salem, Massachusetts, visited Zanzibar in 1846 and stated that the "increasing demand for ivory . . . favor[s] the opinion that before long this valuable and noble animal will be exterminated in this part of Africa." He also mentioned "ivory, borne upon the backs of slaves" in thousands. Reported in Gray, Sir John. (1957) "Trading Expeditions from the Coast to Lakes Tanganyika and Victoria before 1857." *Tanganyika Notes and Records* 49: 226–46.

11. Ivory exports and imports, from which the proportions of the population surviving were calculated, are given in Sheriff, A. (1987) *Slaves, Spices and Ivory in Zanzibar*. Oxford: James Currey; and Spinage, C. A. (1973) "A Review of Ivory Exploitation and Elephant Population Trends in Africa." *East African Wildlife Journal* 11: 281–289.

12. From Kjekshus, H. (1977) *Ecology Control and Economic Development in East Africa*. London: Heinemann Educational Books; as reported on page 25 in Coulson, A. (1982) *Tanzania: A Political Economy*. Oxford: Clarendon Press.

13. Early elephant numbers are given in Grzimek, M., and B. Grzimek. (1960) "A Study of the Game of the Serengeti Plains." *Zeitschrift für Säugetierkunde* 25: 1–61; Stewart, D.R.M. (1962) "Census of Wildlife on the Serengeti, Mara and Loita Plains." *East African Agricultural and Forestry Journal* 28: 58–60; Lamprey, H. F., M. I. M. Turner, and R. H. V. Bell. (1967) "Invasion of the Serengeti National Park by Elephants." *East African Wildlife Journal* 5: 151–66.

14. Elephant numbers are documented in Sinclair, A. R. E., S. A. R.. Mduma, J. G. C. Hopcraft, J. M. Fryxell, R. Hilborn, and S. Thirgood. (2007) "Long-term ecosystem dynamics in the Serengeti: lessons for conservation." *Conservation Biology* 21: 580–590; Morrison, T. A., A. Estes, H. Kija, et al. (2018) "Informing aerial total counts with demographic models: population growth of Serengeti elephants not explained purely by demography. *Conservation Letters* 11: 1–8.

15. In Kruger National Park, continuous culling of elephants from 1965 to 2000 can be treated as experimental removals. Greater reductions resulted in faster returns, meaning that they were density dependent. Similarly, in Hwange National Park, Zimbabwe, elephants were released from culling in 1986, and growth rates slowed as numbers increased due to less available food. See Whyte, I. J., R. van Aarde, and S. L. Pimm. (2003) "Kruger's elephant population: its size and consequences for ecosystem heterogeneity." In J. T. du Toit, K. H. Rogers, and H. C. Biggs (eds.), *The Kruger Experience: Ecology and Management of Savanna Heterogeneity*: Washington, DC: Island Press: 332–348; Chamaille-Jammes, S., H. Fritz, M. Valeix, F. Murindagomo, and J. Clobert. (2008) "Resource variability, aggregation and direct density dependence in an open context: the local regulation of an African elephant population." *Journal of Animal Ecology* 77:135–144; Sinclair, A. R. E., and K. Metzger. (2009) "Advances in wildlife ecology and the influence of Graeme Caughley." *Wildlife Research* 36: 8–15.

16. For bottom-up effects on elephants, see Corfield, T. F. (1973) "Elephant mortality in Tsavo National Park, Kenya." *East African Wildlife Journal* 11: 339–368; Myers, N. (1973) "Tsavo National Park, Kenya, and its elephants: an interim appraisal." *Biological Conservation* 5: 123–132; Laws, R. M., C. Parker, and R. C. B. Johnstone. (1975) *Elephants and Their Habitats*. London: Clarendon Press; Eltringham, S. K. (1980) "A quantitative assessment of range use by large African mammals with particular reference to the effect of elephants on trees." *African Journal of Ecology* 18: 53–71; Owen-Smith, N. (1981) "The white rhino overpopulation problem, and a proposed solution." In J. A. Jewell, S. Holt, and D. Hart (eds.), *Problems in Management of Locally Abundant Wild Mammals*. New York: Academic Press; Owen-Smith, N. (1992) *Megaherbivores: The Influence of Very Large Body Size on Ecology*. Cambridge: Cambridge University Press; and Chamaille-Jammes, S., et al. in note 14 above.

17. Lion population data for 1953 were calculated from transects conducted by Allan Brooks while studying Thomson's gazelle; his unpublished notes were given to me by his widow, Betty, in 2013. For later data see Van Orsdol, K. G., J. P. Hanby, and J. D. Bygott. (1985) "Ecological correlates of lion social organization." *Journal of Zoology London* 206: 97–112; Hanby, J. P., J. D. Bygott, and C. Packer. (1995) "Ecology, demography and behaviour of lions in two contrasting habitats: Ngorongoro Crater and Serengeti plains." In A. R. E. Sinclair and P. Arcese (eds.), *Serengeti II: Dynamics, Management, and Conservation of an Ecosystem*. Chicago: University of Chicago Press: 315–331; Packer, C., R. Hilborn, A. Mosser, et al. (2005) "Ecological change, group territory, and population dynamics in Serengeti lions." *Science* 307: 390–393.

18. Hyena numbers are given in Kruuk, H. (1972) *The Spotted Hyena*. Chicago: University of Chicago Press; Hofer, H., and M. East. (1995) "Population Dynamics, Population Size, and the Commuting System of Serengeti Spotted Hyena." In A. R. E. Sinclair and P. Arcese (eds.), *Serengeti II: Dynamics, Management, and Conservation of an Ecosystem*. Chicago: University of Chicago Press: 332–363.

19. See note 7.

20. For Pleistocene megaherbivores, see Willersley, E., et al. (2014) "Fifty thousand years of Arctic vegetation and megafaunal diet." *Nature* 506: 47–51; Miller, G. H., M. L. Fogel, J. W. Magee, M. K. Gagan, S. J. Clarke, and B. J. Johnson. (2005) "Ecosystem collapse in Pleistocene Australia and a human role in megafaunal extinction." *Science* 309: 287–290.

Chapter 5: How Predators Regulate Prey

1. Sinclair, A. R. E. (1979) "Dynamics of the Serengeti ecosystem. Process and pattern." In A. R. E. Sinclair and M. Norton-Griffiths (eds.), *Serengeti: Dynamics of an Ecosystem*. Chicago: University of Chicago Press: 1–30.

2. Studies on the different types of food each ungulate species prefers showed that each species eats different bits of the plant but that there is a considerable overlap in diet. So we say they have different *niches* but that the overlap can result in competition for food if there is not enough to go around. In the early 1980s, there was considerable debate on the role of competition and predation in determining which species occur and in what numbers. See Strong, D. R., D. Simberloff, L. G. Abele, and A. B. Thistle. (1984) *Ecological Communities: Conceptual Issues and the Evidence*. Princeton: Princeton University Press.

3. The evidence for the effect of predation on shaping the behavior of ungulates is published in Sinclair, A. R. E. (1985) "Does Interspecific Competition or Predation Shape the African Ungulate Community?" *Journal of Animal Ecology* 54: 899–918.

4. To understand the different effects of predators on prey, we must understand a bit about the biology of predators. We can start with the simplest situation, where numbers of prey are very low, having been knocked down by some disturbance (like drought or rinderpest). Initially, prey are hard to find, so predators do not eat many, and prey numbers increase. As prey numbers increase, predators start to search for them, and predator numbers increase because of their increasing food supply. This double effect of eating more and breeding more imposes not just an increase in prey numbers killed but also an increase in the proportion of prey killed, i.e., a density-dependent (regulatory) effect. As with food, the result is a slowing and stabilizing of the prey population. This simple scenario is rarely seen, largely because with only one prey species and one predator species, predator numbers tend to be one step behind those of prey. There is a delay in the response of the predators, which first have to get the extra food before they can breed, and thus are usually a year or so behind. The result is a fluctuating pair of species—first, high numbers of prey, which are then driven down by the subsequent high predator numbers. As prey reach very low numbers, predator populations starve (i.e., they are food regulated), and their numbers collapse until the cycle starts again. We see this in northern Canada with the 10-year cycle of snowshoe hares and their predators, lynx. But if, as we will see for Serengeti in chapter 7, there are many species of prey and predator, then these fluctuations smooth out and a sort of balance is reached. In this scenario the prey numbers are not limited by food.

Under some circumstances—if prey numbers are allowed to increase a long way past this balanced state caused by predators (the equilibrium I talked about in chapters 3 and 8)—a curious phenomenon occurs: prey have lots of food and can breed fast, but the predators cannot eat and breed as fast, so they eat proportionately less and less of the prey. In other words, their impact becomes proportionately less (we call this *inverse density dependence*) so that prey increase faster and faster until eventually they reach the upper limit imposed by their own food supply. What this means is that prey populations can have two equilibria, one set by predators and one set by food.

So far, our discussion has centered around prey that are the main food species (*primary prey*) for the predators. If there are too few prey, predators die off and prey do not go extinct (ignoring other causes of death here). What if the prey species are not the main food but rather just extra, which predators will kill if they can catch them? These *secondary prey* generally are in low numbers to start with (perhaps due to restricted habitat, drought, etc.), and predators tend to eat a set number of these prey irrespective of the size of these prey populations. If the number eaten is greater than the breeding rate of the prey, then of course the secondary prey numbers will go down, and this decrease will progress faster and faster. The proportion of the prey being eaten increases as the population decreases (inverse density dependence) because predator numbers stay the same; thus the prey can be driven to extinction. In summary, predators cannot cause extinction of primary prey but can cause extinction of secondary prey. The theory of predation is outlined in chapter 7 of Fryxell, J., A. R. E. Sinclair, and G. Caughley. (2014) *Wildlife Ecology, Conservation, and Management.* Oxford: Wiley Blackwell, 3rd ed.

5. For regulation and body size, see Sinclair, A. R. E, S. A. R. Mduma, and J. S. Brashares. (2003) "Patterns of predation in a diverse predator-prey system." *Nature* 425: 288–290.

6. The political events that led to the collapse of protection in Tanzania but not Kenya are documented in chapter 9, "Border Closure," of Sinclair, A. R. E. (2012) *Serengeti Story.* Oxford: Oxford University Press.

7. The changes in herbivore numbers with the removal of predators in northern Serengeti are given in note 5 above, and in Sinclair, A. R. E. (1995) "Population limitation of resident herbivores." In A. R. E. Sinclair and P. Arcese (eds.), *Serengeti II: Dynamics, Management, and Conservation of an Ecosystem.* Chicago: University of Chicago Press: 194–219. Buffalo have remained absent in northern Serengeti both during the predator removal and after predators returned. Densities of smaller ungulates were similar in the 1970s, before predator removal and with high buffalo density, and in the 1990s, after predators returned but with no buffalo present. Thus, the absence of buffalo did not explain the observed changes in other ungulates.

8. The role of predators in Kruger National Park is given in Mills, M. G. L., H. C. Biggs, and I. J. Whyte. (1995) "The relationship between rainfall, lion predation and population trends in African herbivores." *Wildlife Research* 22:75–88; Owen-Smith, N., and M. G. L. Mills. (2008) "Predator-prey size relationships in an African large-mammal food web." *Journal of Animal Ecology* 77: 173–183.

9. In North America, see Berger, J., P. B. Stacey, L. Bellis, and M. P. Johnson. (2001) "A mammalian predator-prey imbalance: grizzly bear and wolf extinction affect avian neotropical migrants." *Ecological Applications* 11: 947–960; Ripple, W. J., and R. L. Beschta. (2012) "Trophic cascades in Yellowstone: The first 15 years after wolf reintroduction." *Biological Conservation* 145: 205–213.

10. Georgiadis, N. J., J. G. N. Olwero, G. Ojwang, and S. S. Romanach. (2007) "Savanna herbivore dynamics in a livestock-dominated lands cape: I. Dependence on land use, rainfall, density, and time." *Biological Conservation* 137: 460–472. Georgiadis, N. J., F. Ihwagi, J.G.N. Olwero, and S. S. Romanach. (2007) "Savanna herbivore dynamics in a livestock-dominated landscape. II: Ecological, conservation, and management implications of predator restoration." *Biological Conservation* 137: 473–482.

11. Effect of top predators on other predators can be found in Creel, S. (1992) "Cause of wild dog deaths." *Nature* 360: 633; Laurenson, M. K. (1995) "Implications of high offspring mortality for cheetah population dynamics." In A. R. E. Sinclair and P. Arcese (eds.), *Serengeti II: Dynamics, Management, and Conservation of an Ecosystem.* Chicago: University of Chicago Press: 385–399; Frame, G. W., and L. H. Frame. (1981) *Swift and Enduring: Cheetah and Wild Dogs of the Serengeti.* New York: E. P. Dutton. Also, Palomares, F., and T. M. Caro. (1999) "Interspecific killing among mammalian carnivores." *American Naturalist* 153: 492–508; Swanson, A., T. Caro, H. Davies-Mostert, et al. (2014) "Cheetahs and wild dogs show contrasting patterns of suppression by lions." *Journal of Animal Ecology* 83: 1418–1427; Durant, S. M., M. E. Craft, R. Hilborn, S. Bashir, J. Hando, and L. Thomas. (2011) "Long-term trends in carnivore abundance using distance sampling in Serengeti National Park, Tanzania." *Journal of Applied Ecology* 48: 1490–1500.

12. In conversation with Terry McCabe, who has studied the Ngorongoro Highlands for decades.

13. See note 4.

14. See note 4.

15. Mary Leakey told me that in the 1930s, when she first worked at Olduvai, the eastern plains had long grass. Alan Root, who shot the film *Serengeti Shall Not Die* in the late 1950s, remembered sitting on the front of a Land Rover vehicle and catching harlequin quail with a butterfly net as they flew up in front of the vehicle east of Naabi Hill, where nowadays there is short grass. These quail live only in tall grass (50 cm and higher).

16. Since the late 1960s, this area has supported only short grass some 10 cm high. See also McNaughton, S. J., and F. F. Banyikwa. (1995) "Plant communities and herbivory." In A. R. E. Sinclair and P. Arcese (eds.), *Serengeti II: Dynamics, Management, and Conservation of an Ecosystem.* Chicago: University of Chicago Press: 49–70. See also Olff, H., and M. E. Ritchie. (1998) "Effects of herbivores on grassland plant diversity." *Trends in Ecology and Evolution* 13: 261–265.

17. Anderson, M. T., J. Bukombe, and K. L. Metzger. (2015) "Spatial and temporal drivers of plant structure and diversity in Serengeti savannas." In A. R. E. Sinclair, K. L. Metzger, J. M. Fryxell, and S. A. R. Mduma (eds.), *Serengeti IV: Sustaining Biodiversity in a Coupled Human-Natural System.* Chicago: University of Chicago Press: 105–124.

18. Burning in Serengeti is documented in Norton-Griffiths, M. (1979) "The influence of grazing, browsing, and fire on vegetation dynamics of the Serengeti." In A. R. E. Sinclair and M. Norton-Griffiths (eds.), *Serengeti: Dynamics of an Ecosystem.* Chicago: University of Chicago Press: 310–354. See also Sinclair, A. R. E., S. A. R. Mduma, J. G. C. Hopcraft, J. M. Fryxell, R. Hilborn, and S. Thirgood. (2007) "Long-term ecosystem dynamics in the Serengeti: lessons for conservation." *Conservation Biology* 21: 580–590; Eby, S., J. Dempewolf, R. M. Holdo, and K. L. Metzger. (2015) "Fire in the Serengeti Ecosystem: History, Drivers, and Consequences." In A. R. E. Sinclair, K. L. Metzger, J. M. Fryxell, and S. A. R. Mduma (eds.), *Serengeti IV: Sustaining Biodiversity in a Coupled Human-Natural System.* Chicago: University of Chicago Press: 73–104.

19. Elephant numbers are documented in Sinclair, A. R. E., S. A. R. Mduma, J. G. C. Hopcraft, J. M. Fryxell, R. Hilborn, and S. Thirgood. (2007) "Long-term ecosystem dynamics in the Serengeti: lessons for conservation." *Conservation Biology* 21: 580–590; Morrison, T. A., A. Estes, H. Kija, H. Maliti, S. A. R. Mduma, A. R. E. Sinclair, and E. Kohi. (2017) "Informing aerial total counts with demographic models: population growth of Serengeti elephants not explained purely by demography." *Conservation Letters* 11: 1–8.

20. For indirect effects of large herbivores on other fauna, see Nkwabi, A. K., K. Metzger, R. Beyers, F. Magige, S. A. R. Mduma, J. G. C. Hopcraft, and A. R. E. Sinclair. (2019) "Bird community responses to changes in vegetation caused by increasing large mammal populations in the Serengeti woodlands." *Wildlife Research* 46: 256–264. In Kenya, see Long, R. A., A. Wambua, J. R. Goheen, T. M. Palmer, and R. M. Pringle. (2017) "Climatic variation mediates the indirect effects of large herbivores on small-mammal habitat use." *Journal of Animal Ecology* 86: 739–748; Pringle, R. M., R. L. Sensenig, D. Kimuyu, et al. (2015) "Synergistic indirect effects of fire and elephants on arboreal fauna." *Journal of Animal Ecology* 84: 1637–1645; Coverdale, T. C., T. R. Kartzinel, K. L. Grabowski, et al. (2016) "Elephants in the understory: opposing direct and indirect effects of consumption and ecosystem engineering by megaherbivores on African sa-

vanna plants." *Ecology* 97: 3219–3230; Ogada, D. L., M. E. Gadd, R. S. Ostfeldt, T. P. Young, and F. Keesing. (2008) "Impacts of large herbivorous mammals on bird diversity and abundance in an African savanna." *Oecologia* 156: 387–397; Herremans, M. (1995) "Effects of woodland modification by African elephant *Loxodonta africana* on bird diversity in northern Botswana." *Ecography* 18: 440–454.

21. There is a vast literature on the effect of herbivores on plants. A few references for Kruger National Park, South Africa, are Scholes, R. J., W. J. Bond, and H. C. Eckhardt. (2003) "Vegetation dynamics in the Kruger ecosystem." In J. T. du Toit, K. H. Rogers, and H. C. Biggs (eds.), *The Kruger Experience: Ecology and Management of Savanna Heterogeneity*. Washington, DC: Island Press: 242–262. For southern Africa, see Fritz, H., P. Duncan, I. J. Gordon, and A. W. Illius. (2002) "Megaherbivores influence trophic guilds structure in African ungulate communities." *Oecologia* 131: 620–625. For northern Kenya, see Pringle, R. M., T. P. Young, D. I. Rubenstein, and D. J. McCauley. (2007) "Herbivore-initiated interaction cascades and their modulation by productivity in an African savanna." *Proceedings of the National Academy of Sciences, USA* 104: 193–197. For Yellowstone National Park, USA, see Beschta, R. L., and W. J. Ripple. (2010) "Recovering riparian plant communities in northern Yellowstone, USA." *Restoration Ecology* 18: 380–389; White, P. J., R. A. Garrott, and G. E. Plumb (eds.). (2013) *Yellowstone's Wildlife in Transition*. Cambridge, MA: Harvard University Press.

22. See note 21 and Daskin, J. H., and R. M. Pringle. (2016) "Does primary productivity modulate the indirect effects of large herbivores? A global meta-analysis." *Journal of Animal Ecology* 85: 857–868.

23. Nkwabi, A. K., A. R. E. Sinclair, K. L. Metzger, and S. A. R. Mduma. (2011) "Disturbance, ecosystem function and compensation: the influence of wildfire and grazing on the avian community in the Serengeti ecosystem, Tanzania." *Austral Ecology* 36: 403–412; Nkwabi, A. K., A. R. E. Sinclair, and S. A. R. Mduma. (2015) "The effect of natural disturbances on the avian community of the Serengeti woodlands." In A. R. E. Sinclair, K. L. Metzger, J. M. Fryxell, and S. A. R. Mduma (eds.), *Serengeti IV: Sustaining Biodiversity in a Coupled Human-Natural System*. Chicago: University of Chicago Press: 395–418. Elsewhere, see Pringle, R. P., D. M. Kimuyu, R. L. Sensenig, et al. (2015) "Synergistic effects of fire and elephants on arboreal animals in an African savanna." *Journal of Animal Ecology* 84: 1637–1645.

24. For effects of giraffe, see Strauss, M. K. L., M. Kilewo, D. Rentsch, and C. Packer. (2015) "Food supply and poaching limit giraffe abundance in the Serengeti." *Population Ecology* 57: 505–516; Bond, W. J., and D. Loffell. (2001) "Introduction of giraffe changes acacia distribution in a South African savanna." *African Journal of Ecology* 39: 286–294; Staver, A. C., and W. J. Bond. (2014) "Is there a 'browse trap'? Dynamics of herbivore impacts on trees and grasses in an African savanna." *Journal of Ecology* 102: 595–602, 891.

25. For top-down effects of megaherbivores, see Laws, R. M. (1970) "Elephants as agents of habitat and landscape change in East Africa." *Oikos* 21: 1–15; Norton-Griffiths, M. (1979) "The influence of grazing, browsing, and fire on vegetation dynamics of the Serengeti." In A. R. E. Sinclair and M. Norton-Griffiths (eds.), *Serengeti: Dynamics of an Ecosystem*. Chicago: University of Chicago Press: 310–354; Eltringham, S. K. (1980) "A quantitative assessment of range use by large African mammals with particular reference to the effect of elephants on trees." *African Journal of Ecology* 18: 53–71; Owen-Smith, N. (1988) *Megaherbivores: The Influence of Very Large*

Body Size on Ecology. Cambridge: Cambridge University Press; Eckhardt, H. C., B. W. van Wilgen, and H. C. Biggs. (2000) "Trends in woody vegetation cover in the Kruger National Park, South Africa, between 1940 and 1998." *African Journal of Ecology* 38: 108–115. See also note 21.

26. See note 29 for top-down effects of elephants. For bottom-up effects on elephants, see Corfield, T. F. (1973) "Elephant mortality in Tsavo National Park, Kenya." *East African Wildlife Journal* 11: 339–368; Croze, H., A. K. K. Hillman, and E. M. Lang. (1981) "Elephants and their habitats: how do they tolerate each other?" In C. W. Fowler and T. D. Smith (eds.), *Dynamics of Large Mammal Populations.* New York: John Wiley & Sons: 297–316; Owen-Smith, N. (1981) "The white rhino overpopulation problem, and a proposed solution." In J. A. Jewell, S. Holt, and D. Hart (eds.), *Problems in Management of Locally Abundant Wild Mammals.* New York: Academic Press: 129–150; Chamaille-Jammes, S., H. Fritz, M. Valeix, F. Murindagomo, and J. Clobert. (2008) "Resource variability, aggregation and direct density dependence in an open context: the local regulation of an African elephant population." *Journal of Animal Ecology* 77: 135–144.

27. Moe, S. R., L. P. Rutina, H. Hyttenborn, and J. T. du Toit. (2009) "What controls woodland regeneration after elephants have killed big trees?" *Journal of Applied Ecology* 46: 223–230.

28. See note 26. Additional information on the extirpation of elephants during the ivory trade can be found in Patterson, J. H. (1907). *The Maneaters of Tsavo.* London: Fontana; Skarpe, C., P. A. Aarrestad, H. P. Andreassen, et al. (2004) "The return of the giants: Ecological effects of an increasing elephant population." *Ambio* 33: 276–282.

29. Hopcraft, J. G. C., H. Olff, and A. R. E. Sinclair. (2010) "Herbivores, resources, and risks: Alternating regulation along primary environmental gradients in savannas." *Trends in Ecology & Evolution* 25: 119–128.

30. For a synthesis of trophic cascades, see Terborgh, J., and J. A. Estes. (2010) *Trophic Cascades: Predators, Prey, and the Changing Dynamics of Nature.* Washington, DC: Island Press; Estes, J. A., J. Terborgh, J. S. Brashares, A. R. E. Sinclair, et al. (2011) "Trophic downgrading of planet earth." *Science* 333: 301–306.

Chapter 6: How Migration Structures Serengeti

1. Percival, A. B. (1924) *A Game Ranger's Note Book.* London: Nisbet and Co.; Percival, A. B. (1928) *A Game Ranger on Safari.* London: Nisbet & Co.

2. For the early work on the wildebeest migration, see Grzimek, B., and M. Grzimek. (1960) *Serengeti Shall Not Die.* London: Hamish Hamilton, Ltd. (documentary film); Grzimek, M., and B. Grzimek. (1960) "A Study of the Game of the Serengeti Plains." *Zeitschrift für Säugetierkunde* 25: 1–61; Stewart, D. R. M. (1962) "Census of Wildlife on the Serengeti, Mara and Loita Plains." *East African Agricultural and Forestry Journal* 28: 58–60; Talbot, L. M., and M. H. Talbot. (1963) "The wildebeest in western Masailand." *Wildlife Monographs No. 12.* Washington, DC: The Wildlife Society; Watson, R. M. (1967) "The Population Ecology of the Wildebeest (*Connochaetes taurinus albojubatus*) in the Serengeti." PhD dissertation. University of Cambridge.

3. Linda Maddock provided the first quantitative description of the migration. See Maddock, L. (1979) "The 'Migration' and Grazing Succession." In A. R. E. Sinclair and M. Norton-Griffiths (eds.), *Serengeti: Dynamics of an Ecosystem.* Chicago: University of Chicago Press: 104–129.

4. Lion populations on the plains have been studied continuously since George Schaller began (1966–69), followed by Brian Bertram (1969–73), Jeanette Hanby and David Bygott (1974–77), and Craig Packer, who continued for over two decades (1978–2014). Research in the 2020s is under two programs directed by Mike Anderson of Wake Forest University and Jerry Belant of Mississipi State University. See Schaller, G. B. (1972) *The Serengeti Lion*. Chicago: University of Chicago Press; Bertram, B. C. R. (1979) "Serengeti predators and their social systems." In A. R. E. Sinclair and M. Norton-Griffiths (eds.), *Serengeti: Dynamics of an Ecosystem*. Chicago: University of Chicago Press: 221–248; Hanby, J. P., and J. D. Bygott. (1995) "Population changes in lions and other predators." In A. R. E. Sinclair and M. Norton-Griffiths (eds.), *Serengeti: Dynamics of an Ecosystem*. Chicago: University of Chicago Press: 249–262; Packer, C., R. Hilborn, A. Mosser, et al. (2005) "Ecological change, group territoriality and population dynamics in Serengeti lions." *Science* 307: 390–393.

5. For zebra behavior see Sinclair, A. R. E. (1985) "Does interspecific competition or predation shape the African ungulate community?" *Journal of Animal Ecology* 54: 899–918.

6. See Metzger, K. L., A. R. E. Sinclair, A. Macfarlane, M. B. Coughenour, and J. Ding. (2015). "Scales of change in the Greater Serengeti ecosystem." In A. R. E. Sinclair, K. L. Metzger, J. M. Fryxell, and S. A. R. Mduma (eds.), *Serengeti IV: Sustaining Biodiversity in a Coupled Human-Natural System*. Chicago: University of Chicago Press: 33–72.

7. In Huxley, J. (1936) *Africa View*. London: Chatto and Windus.

8. See note 3.

9. Hubert Braun worked on the grasslands of Serengeti 1966–69, funded by The Netherlands Tropical Research Institute (WOTRO). See Braun, H. M. H. (1973) "Primary production in the Serengeti: purpose, methods and some results of research." *Annals of the University d'Abidjan (E)* 6:171–188.

10. For nutrients on the plains and migration, see McNaughton, S. J. (1990) "Mineral nutrition and seasonal movements of African migratory ungulates." *Nature* 345: 613–615; Murray, M. G. (1995) "Specific nutrient requirements and migration of wildebeest." In A. R. E. Sinclair and P. Arcese (eds.), *Serengeti II: Dynamics, Management, and Conservation of an Ecosystem*. Chicago: University of Chicago Press: 231–256: Fryxell, J. M. "Aggregation and migration by grazing ungulates in relation to resources and predators." In A. R. E. Sinclair and P. Arcese (eds.), *Serengeti II: Dynamics, Management, and Conservation of an Ecosystem*. Chicago: University of Chicago Press: 257–273.

11. The role of termites in soil nutrition is described in Freymann, B. P., R. Buitenwerf, O. Desouza, and H. Olff. (2008) "The importance of termites (*Isoptera*) for the recycling of herbivore dung in tropical ecosystems: a review." *European Journal of Entomology* 105: 165–173; Bonachela, J. A., R. M. Pringle, E. Sheffer, et al. (2015) "Termite mounds can increase the robustness of dryland ecosystems to climatic change." *Science* 347: 651–65.

12. For adaptations of migration, see Sinclair, A. R. E. (1983) "The function of distance movements in vertebrates." In I. R. Swingland and P. J. Greenwood (eds.), *The Ecology of Animal Movement*. Oxford: Oxford University Press: 240–258. See also Milner-Gulland, E. J., J. M. Fryxell, and A. R. E. Sinclair (eds.). (2011) *Animal Migration: A Synthesis* Oxford: Oxford University Press.

13. John Fryxell conducted his PhD work in Boma, southern Sudan, from 1980 to 1983, at which point he narrowly escaped capture by Sudanese rebels. He joined the faculty of the

University of Guelph in 1986 and became head of the Department of Integrative Biology in 2014. In 2018 he became executive director of the Biodiversity Institute of Ontario (BIO). He has been involved with Serengeti work since 1991 and has worked with the Serengeti Biodiversity Program since 2001.

14. For John Fryxell's work on kob migrations, see Fryxell, J. M. (1987) "Seasonal reproduction of white-eared kob in Boma National Park, Sudan." *African Journal of Ecology* 25: 117–124; Fryxell, J. M. (1987) "Food limitation and demography of a migratory antelope, the white-eared kob." *Oecologia* 72: 83–91; Fryxell, J. M., and A. R. E. Sinclair. (1988) "Seasonal migration by white-eared kob in relation to resources." *African Journal of Ecology.* 26: 17–31.

15. Fryxell, J. M., and A. R. E. Sinclair. (1988) "Causes and consequences of migration by large herbivores." *Trends in Ecology and Evolution* 3: 237–241.

16. See note 12.

17. For wildebeest energetics, see Pennycuick, C. J. (1979) "Energy costs of locomotion and the concept of 'foraging radius.'" In A. R. E. Sinclair and M. Norton-Griffiths (eds.), *Serengeti: Dynamics of an Ecosystem.* Chicago: University of Chicago Press: 164–184.

18. Holdo, R. M., R. D. Holt, M. B. Coughenour, and M. E. Ritchie. (2007) "Plant productivity and soil nitrogen as a function of grazing, migration and fire in an African savanna." *Journal of Ecology* 95: 115–128.

19. See Kruuk, H. (1972) *The Spotted Hyena.* Chicago: University of Chicago Press.

20. See note 4.

Chapter 7: Biodiversity and Stability of Ecosystems

1. The United Nations Conference on Environment and Development was held in Rio de Janeiro, Brazil, in 1992. Termed the Earth Summit, it set out a list of principles for international development and environmental cooperation, and in particular the Convention on Biodiversity, signed by 152 countries.

2. Hugh Lamprey documented for the first time the habitat partitioning of ungulates in East Africa, focusing his studies on Tarangire National Park. Lamprey, H. F. (1963) "Ecological Separation of the Large Mammals in the Tarangire Game Reserve, Tanganyika." *East African Wildlife Journal* 1: 63–92. For southern Africa see Ferrar, A. A., and B. H. Walker. (1974) "An Analysis of Herbivore/Habitat Relationships in Kyle National Park, Rhodesia." *Journal of the Southern African Wildlife Management Association* 4: 137–147; du Toit, J. T. (1990) "Feeding-height stratification among African browsing ruminants." *African Journal of Ecology* 28: 55–82.

3. For diet niches of African ungulates, see Jarman, P. J. (1971) "Diets of large mammals in the woodlands around Lake Kariba, Rhodesia." *Oecologia* 8: 157–178; Gwynne, M. D., and R. H. V. Bell. (1968) "Selection of Vegetation Components by Grazing Ungulates in the Serengeti National Park." *Nature* 220: 390–393; Murray, M. G. (1993) "Comparative Nutrition of Wildebeest, Hartebeest, and Topi in the Serengeti." *African Journal of Ecology* 31: 172–177; Murray, M. G., and D. Brown. (1993) "Niche Separation of Grazing Ungulates in the Serengeti: An Experimental Test." *Journal of Animal Ecology* 62: 380–389. See also Kartinzel, T. R., P. Chen, T. C. Coverdale, et al. (2015) "DNA metabarcoding illuminates dietary niche partitioning by large African herbivores." *Proceedings of the National Academy of Sciences, USA* 112: 8019–8024.

For diet overlap of ungulates in East Africa, see Sinclair, A. R. E. (1985) "Does Interspecific Competition or Predation Shape the African Ungulate Community?" *Journal of Animal Ecology* 54: 899–918. For southern Africa, see Kleynhans, E. J., A. E. Jolles, M. R. E. Bos, and H. Olff. (2011) "Resource partitioning along multiple niche dimensions in differently sized African savanna grazers." *Oikos* 120: 591–600.

For skull and mouth adaptations, see Codron, D., J. S. Brink, L. Rossouw, et al. (2008) "Functional Differentiation of African Grazing Ruminants: An Example of Specialized Adaptations to Very Small Changes in Diet." *Biological Journal of the Linnean Society* 94: 755–764; Janis, C. M., and D. Ehrhardt. (1988) "Correlation of Relative Muzzle Width and Relative Incisor Width with Dietary Preference in Ungulates." *Zoological Journal of the Linnean Society* 92: 267–84; Gordon, I. J., and A. W. Illius. (1996) "The Nutritional Ecology of African Ruminants: A Reinterpretation." *Journal of Animal Ecology* 65: 18–28; Gagnon, M., and A. E. Chew. (2000) "Dietary Preferences in Extant African Bovidae." *Journal of Mammalogy* 81: 490–511.

4. Modern analyses have confirmed earlier findings on the habitat differences and overlaps of the ungulates in Serengeti. See Anderson, T. M, S. White, B. Davis, et al. (2016) "The spatial distribution of African savannah herbivores: species associations and habitat occupancy in a landscape context." *Philosophical Transactions of the Royal Society, B.* 371: 20150314.

5. Our conclusions that competition was the main driver in the ecosystem are presented in Sinclair, A.R.E. (1979) "Dynamics of the Serengeti Ecosystem: Process and Pattern." In A. R. E. Sinclair and M. Norton-Griffiths (eds.), *Serengeti: Dynamics of an Ecosystem.* Chicago: University of Chicago Press: 1–30.

6. The keystone concept can be found in Paine, R. T. (1966) "Food web complexity and species diversity." *American Naturalist* 100: 65–75; Paine, R. T. (1980) "Food webs: linkage, interaction strength and community infrastructure." *Journal of Animal Ecology* 49: 667–685. In 1977, Bob Holt of the University of Florida showed how predators could keep prey populations at low enough levels that they did not compete with each other. He called this theory *apparent competition*, which later became known as *predator-mediated coexistence.* See Holt, R. D. (1977) "Predation, apparent competition, and the structure of prey communities." *Theoretical Population Biology* 12: 197–229.

7. See Estes, J. A., and D. O. Duggins. (1995) "Sea otters and kelp forests in Alaska: generality and variation in a community ecology paradigm." *Ecological Monographs* 65: 75–100.

8. In the early 1980s there was considerable debate on the role of competition and predation in determining which species occur and in what numbers. See Strong, D. R., D. Simberloff, L. G. Abele, and A. B. Thistle. (1984) *Ecological Communities: Conceptual Issues and the Evidence.* Princeton: Princeton University Press. For a theory of how top-down processes shape species diversity, see Terborgh, J. W. (2015) "Toward a trophic theory of species diversity." *Proceedings of the National Academy of Sciences, USA* 112: 11415–11422.

9. Peter Jarman studied the social organization of ungulates at Lake Kariba, Zimbabwe: Jarman, P. J. (1974) "The Social Organization of Antelope in Relation to Their Ecology." *Behaviour* 48: 215–266. He came to Serengeti to study impala 1968–71.

10. In the 1990s, a debate developed among scientists as to whether the abundance of biological species in an ecosystem contributed to long-term stability of that system. It was suggested that the many species living in an ecosystem provided some sort of backup in case some

species were lost because of accidents or outside disturbances. Many species are very similar in their role—their use of food and space, as well as the predator species they support—so that if one species disappears due to a disturbance, another species can take over the role. In this way the whole community of species can continue much as before, provided, of course, that not too many species are lost. If the disturbance is huge, large numbers of species are extirpated, and there are not enough left to fill the vacant niches, causing the ecosystem to unravel. This was the theory—small disturbances can be buffered by the large diversity of species present, but large disturbances cannot because too many species are lost. It was a simple and elegant theory at face value and comforting if it could be shown to be correct. Unfortunately, difficulties became apparent as soon as scientists tried to test it. First, species are all different and none can take over another's niche exactly. Second, the buffering effect really depends on which species are lost, because some are more important than others in the workings of the ecosystem. For an overview of some of these issues, see Walker, B. H. (1992) "Biological Diversity and Ecological Redundancy." *Conservation Biology* 6: 18–23; Walker, B. H. (1995) "Conserving Biological Diversity through Ecosystem Resilience." *Conservation Biology* 9: 747–752; Kinzig, A. P., S. W. Pacala, and D. Tilman. (eds.) (2001) *The Functional Consequences of Biodiversity*. Princeton: Princeton University Press; Power, M. E., D. Tilman, J. A. Estes, et al. (1996) "Challenges in the Quest for Keystones." *BioScience* 46: 609–620.

11. Nkwabi, A. K., A. R. E. Sinclair, K. L. Metzger, and S. A. R. Mduma. (2011) "Disturbance, ecosystem function and compensation: the influence of wildfire and grazing on the avian community in the Serengeti ecosystem, Tanzania." *Austral Ecology* 36: 403–412.

12. Nkwabi, A. K., K. Metzger, R. Beyers, F. Magige, S. A. R. Mduma, J. G. C. Hopcraft, and A. R. E. Sinclair. (2019) "Bird community responses to changes in vegetation caused by increasing large mammal populations in the Serengeti woodlands." *Wildlife Research* 46: 256–264.

13. Variability of numbers is seen in figure 9.2 of Sinclair, A. R. E. (1995) "Population limitation of resident herbivores." In A. R. E. Sinclair and P. Arcese (eds.), *Serengeti II: Dynamics, Management, and Conservation of an Ecosystem*. Chicago: University of Chicago Press: 194–219.

14. See note 11.

15. More biodiversity within and among trophic levels resulted in higher regulation (and lower variability) of individual species populations and more stability of the community as a whole. This has also been observed in other systems: for example, in predator-prey systems with fish in coral reefs; see Carr, M. H., T. W. Anderson, and M. A. Hixon. (2002) "Biodiversity, Population Regulation, and the Stability of Coral-Reef Fish Communities." *Proceedings of the National Academy of Sciences, USA* 99 (17): 11241–11245. In other systems increased species diversity within a trophic level led to more stability of the community as a whole but increased variability between populations of species, as in some plant communities. See Tilman, D. (1996) "Biodiversity: Population versus Ecosystem Stability." *Ecology* 77:350–363.

16. Peters, C. R., R. J. Blumenschine, R. L. Hay, et al. (2008) "Paleoecology of the Serengeti-Mara ecosystem." In A. R. E. Sinclair, C. Packer, S. A. R. Mduma, and J. M. Fryxell (eds.), *Serengeti III: Human Impacts on Ecosystem Dynamics*. Chicago: University of Chicago Press: 47–94.

17. Food limitation in Kruger Park, South Africa, is documented in Owen-Smith, N. (1990) "Demography of a large herbivore, the greater kudu *Tragelaphus strepsiceros*, in relation to rainfall." *Journal of Animal Ecology* 59: 893–913; Owen-Smith, N., and J. O. Ogutu. (2003) "Rainfall

influences on ungulate population dynamics in the Kruger National Park." In J. T. du Toit, K. H. Rogers, and H. C. Biggs (eds.), *The Kruger Experience*. Washington, DC: Island Press: 310–331.

18. For the consequences of body size, see Sinclair, A. R. E., S. A. R. Mduma, and J. S. Brashares. (2003) "Patterns of predation in a diverse predator-prey system." *Nature* 425: 288–290. In Kruger Park, see du Toit, J. T., and N. Owen-Smith. (1989) "Body Size, Population Metabolism, and Habitat Specialization among Large African Herbivores." *American Naturalist* 133: 736–740; Radloff, F. G. T., and J. T. du Toit. (2004) "Large predators and their prey in a southern African savanna: a predator's size determines its prey size range." *Journal of Animal Ecology* 73: 410–423.

19. Brashares, J. S., L. R. Prugh, C. J. Stoner, and C. W. Epps. (2010) "Mesopredator release." In J. Terborgh and J. A. Estes (eds.), *Trophic Cascades: Predators, Prey, and the Changing Dynamics of Nature*. Washington, DC: Island Press: 221–240.

20. Effects of wolf removals are presented in White, P. J., R. A. Garrott, and G. E. Plumb (eds.). (2013) *Yellowstone's Wildlife in Transition*. Cambridge, MA.: Harvard University Press.

21. Loss of large carnivores would result in mesopredator release (the increase in medium-sized carnivores) so that there would be an outbreak of larger antelopes like kongoni but a decrease of smaller ones like impala. Loss of small carnivores would result in an outbreak of small herbivores. On the prey side, loss of large prey that are eaten by carnivores, such as buffalo and wildebeest, would result in a decline of large predators. Loss of small prey would result in loss of small predators. We have some evidence of these changes during the rinderpest era, when lions and hyena were much scarcer. No matter what component is lost, we see changes that could destabilize the community of mammals, with reverberations felt through other parts of the ecosystem, such as birds and plants.

22. For termites in Serengeti, see Freyman, B. P., S. N. de Visser, and H. Olff. (2010) "Spatial and temporal hotspots of termite-driven decomposition in the Serengeti." *Ecography* 33: 443–450. For aardwolf, see Mills, G., and M. Harvey. (2001) *African Predators*. Washington, DC: Smithsonian Institution Press; Kruuk, H., and W. A. Sands. (1972) "The aardwolf (*Proteles cristatus, Sparrman*) as predator of termites." *East African Wildlife Journal* 10: 211–227.

23. Desmond Vesey-Fitzgerald was one of the great old-fashioned naturalists. From 1933 to 1974, he worked in Brazil, British Guiana, the British West Indies, the Seychelles, Madagascar, coastal East Africa, Malaya, Sudan, Saudi Arabia, Oman, and Kenya, finally ending up as park warden of Arusha National Park in Tanzania. But his greatest work was the first description of the keystone effect of elephants on grazing ecosystems in East Africa. See Vesey-Fitzgerald, D. F. (1960) "Grazing Succession among East African Game Animals." *Journal of Mammalogy* 41: 161–172.

24. See Bell, R. H. V. (1970) "The Use of the Herb Layer by Grazing Ungulates in the Serengeti." In A. Watson (ed.), *Animal Populations in Relation to Their Food Resources*. Oxford: Blackwell Scientific: 111–123; Bell, R. H. V. (1971) "A Grazing Ecosystem in the Serengeti." *Scientific American* 224: 86–93.

25. There is some debate on when and how compensatory growth of grazed plants takes place. See Belsky, A. J. (1986) "Does herbivory benefit plants?" *American Naturalist* 127, 777–783; McNaughton, S. J. (1976) "Serengeti migratory wildebeest: facilitation of energy flow by grazing." *Science* 191: 92–94. McNaughton, S. J. (1983) "Compensatory plant growth as a response to herbivory." *Oikos* 40: 329–336.

26. Gazelle numbers are given in Dublin, H. T., A. R. E. Sinclair, S. Boutin, E. Anderson, M. Jago, and P. Arcese. (1990) "Does competition regulate ungulate populations? Further evidence from Serengeti, Tanzania." *Oecologia* 82: 238–288.

27. For commensal interactions between plants, see McNaughton, S. J. (1978) "Serengeti ungulates: feeding selectivity influences the effectiveness of plant defense guilds." *Science* 199: 806–807. See also Louthan, A. M., D. F. Doak, J. R. Goheen, T. M. Palmer, and R. M. Pringle. (2014) "Mechanisms of plant-plant interactions: concealment from herbivores is more important than abiotic stress mediation in an African savannah." *Proceedings of the Royal Society B 281 (1780). Article Number: 20132647*; Stachowicz, J. J. (2001) "Mutualism, Facilitation, and the Structure of Ecological Communities." *BioScience* 51: 235–246.

28. The candelabrum tree (*Euphorbia candelabrum*) is a succulent that grows into a tree up to 12 m high and with a bole diameter of 50 cm or more. Anna Sinclair noticed that in the general savanna, where fire is frequent, young plants are only found within the dense vegetation at the base of regenerating umbrella acacia some 3 m high, as a protection from fire.

29. For effects of trees on soil and grass layer, see Belsky, A. J. (1994) "Influences of trees on savanna productivity: test of shade, nutrients, and tree-grass competition." *Ecology* 75: 922–932; Treydte, A. C., I. M. A. Heitkonig, H. H. T. Prins, and F. Ludwig. (2007) "Trees Improve Grass Quality for Herbivores in African Savannas." *Perspectives in Plant Ecology Evolution and Systematics* 8: 197–205.

30. The relationship between Fischer's lovebirds and rufous-tailed weavers is given in Mwangomo, E. A., L. H. Hardesty, A. R. E. Sinclair, S. A. R. Mduma, and K. Metzger. (2007) "The association of two endemic birds in the Serengeti ecosystem: habitat selection, diet and group formation of the Rufous-tailed weaver, Fischer's lovebird and associated species. *African Journal of Ecology* 46: 267–275.

31. Mycorrhizal associations with grass have been highlighted in the recent analysis of the wildebeest migration and their grazing on granitic and volcanic soils. See Veldhuis, M. P., M. E. Ritchie, J. O. Ogutu, et al. (2019) "Cross-boundary human impacts compromise the Serengeti-Mara ecosystem." *Science* 363: 1424–1428.

32. See Palmer, M. S., and C. Packer. (2018) "Giraffe bed and breakfast: Camera traps reveal Tanzanian yellow-billed oxpeckers roosting on their large mammalian hosts." *African Journal of Ecology* 56: 882–884.

33. Hunting parties of insectivorous birds in the tops of African acacia trees include black-bellied apalis (*Apalis flavida*), banded tit-flycatcher (*Sylvia boehmi*), red-faced crombec (*Sylvietta whytii*), buff-bellied warbler (*Phyllolais pulchella*), red-throated tit (*Melaniparus fringillinus*), and chin-spot flycatcher (*Batis molitor*).

34. Goheen, J. R., and T. M. Palmer. (2010) "Defensive plant-ants stabilize megaherbivore-driven landscape change in an African savanna." *Current Biology* 20: 1768–1772.

35. For the stability of predator-prey systems, see Fryxell, J., A. Mosser, A. R. E. Sinclair, and C. Packer. (2007) "Group Formation and Predator–Prey Dynamics in Serengeti." *Nature* 449: 1041–1044.

36. Hale, K. R. S., F. S. Valdovinos, and N. D. Martinez. (2020) "Mutualism increases diversity, stability, and function of multiplex networks that integrate pollinators into food webs." *Nature Communications* 11 (2182): Article Number: 2182.

Chapter 8: Disturbance and the Persistence of Ecosystems

1. Kristine Metzger, who has the wonderful ability to work in remote areas of Serengeti on plant communities while also being responsible for the management of all our data since 1998, has put together both the environments of the distant past and the present-day changes in disturbances due to rainfall and fire. See Metzger, K. L., A. R. E. Sinclair, A. Macfarlane, M. B. Coughenour, and J. Ding. (2015) "Scales of change in the Greater Serengeti ecosystem." In A. R. E. Sinclair, K. L. Metzger, J. M. Fryxell, and S. A. R. Mduma (eds.), *Serengeti IV: Sustaining Biodiversity in a Coupled Human-Natural System.* Chicago: University of Chicago Press: 33–72.

2. See note 1.

3. Southern Oscillation Index data are from Australian Government Bureau of Meteorology. http://www.bom.gov.au/climate/enso/soi/.

4. Ogutu, J. O., and N. Owen-Smith. (2003) "ENSO, rainfall and temperature influences on extreme population declines among African savanna ungulates." *Ecology Letters* 6:412–419; Ogutu, J. O., H. P. Piepho, H. T. Dublin, N. Bhola, and R. S. Reid. (2008) "El Niño-Southern Oscillation, Rainfall, Temperature and Normalized Difference Vegetation Index Fluctuations in the Mara-Serengeti ecosystem." *African Journal of Ecology* 46: 132–143.

5. The effects of El Niño are described in Sinclair, A. R. E., K. L. Metzger, J. M. Fryxell, et al. (2013) "Asynchronous food web pathways buffer the response of Serengeti predators to El Niño Southern Oscillation." *Ecology* 94: 1123–1130.

6. See note 5.

7. For butterflies, see Sinclair, A. R. E., A. K. Nkwabi, and K. L. Metzger. (2015) "The butterflies of Serengeti: Impact of environmental disturbance on biodiversity." In A. R. E. Sinclair, K. L. Metzger, J. M. Fryxell, and S. A. R. Mduma (eds.), *Serengeti IV: Sustaining Biodiversity in a Coupled Human-Natural System.* Chicago: University of Chicago Press: 301–322.

8. See note 5.

9. For detailed descriptions of the migration movements, see chapter 6, and Hopcraft, J. G. C., R. M. Holdo, E. Mwangomo, et al. (2015) "Why are wildebeest the most abundant herbivore in the Serengeti ecosystem?" In A. R. E. Sinclair, K. L. Metzger, J. M. Fryxell, and S. A. R. Mduma (eds.), *Serengeti IV: Sustaining Biodiversity in a Coupled Human-Natural System.* Chicago: University of Chicago Press: 125–174.

10. Nkwabi, A. K. (2016) "Influences of Habitat Structure and Seasonal Variation on Abundance, Diversity and Breeding of Bird Communities in Selected Parts of the Serengeti National Park, Tanzania." PhD thesis. Dar es Salaam: University of Dar es Salaam.

11. Guineafowl (*Numida meleagris*) are ground birds related to pheasants. White-browed coucal (*Centropus superciliosus*) is one of the nonparasitic cuckoo group found in Africa, Asia, and Australia. Both nest on the ground in long, rank grass.

12. The army worm moth (*Spodoptera exigua*) comes in with the first rainstorms in November. For insect movements, see Sinclair, A. R. E. (1978) "Factors affecting the food supply and breeding season of resident birds and movements of Palaearctic migrants in a tropical African savannah." *Ibis* 120: 480–497.

13. Early studies of fire are Norton-Griffiths, M. (1979) "The influence of grazing, browsing, and fire on the vegetation dynamics of the Serengeti." In A.R.E. Sinclair and M. Norton-Griffiths

(eds.), *Serengeti: Dynamics of an Ecosystem.* Chicago: University of Chicago Press: 310–352; Stronach, N.R.H., and S. J. McNaughton. (1989) "Grassland fire dynamics in the Serengeti ecosystem, and a potential method of retrospectively estimating fire energy." *Journal of Applied Ecology* 26: 1025–1033. The modern synthesis of fire is presented by Eby, S., J. Dempewolf, R. Holdo, and K. Metzger. (2015) "Fire in the Serengeti Ecosystem: History, Drivers, and Consequences." In A. R. E. Sinclair, K. L. Metzger, J. M. Fryxell, and S. A. R. Mduma (eds.), *Serengeti IV: Sustaining Biodiversity in a Coupled Human-Natural System.* Chicago: University of Chicago Press: 73–104.

14. Holdo, R. M., R. D. Holt, and J. M. Fryxell. (2009) "Opposing rainfall and plant nutritional gradients best explain the wildebeest migration in the Serengeti." *Amercian Naturalist* 173: 431–445; Holdo, R. M., R. D. Holt, and J. M. Fryxell. (2009) "Grazers, browsers, and fire influence the extent and spatial pattern of tree cover in the Serengeti." *Ecological Applications* 19: 95–109; Holdo, R. M., A. R. E. Sinclair, K. L. Metzger, et al. (2009) "A disease-mediated trophic cascade in the Serengeti and its implications for ecosystem C." *PloS Biology* 7: e1000210.

15. See note 13.

16. Grazing by domestic cattle due to human encroachment is documented in Veldhuis, M. P., M. E. Ritchie, J. O. Ogutu, et al. (2019) "Cross-boundary human impacts compromise the Serengeti-Mara ecosystem." *Science* 363: 1424–1428.

17. Ford, J. (1971) *The Role of Trypanosomiases in African Ecology.* Oxford: Clarendon Press.

18. The decline of tree numbers in Serengeti is documented in Norton-Griffiths, M. (1979) "The influence of grazing, browsing, and fire on the vegetation dynamics of the Serengeti." In A. R. E. Sinclair and M. Norton-Griffiths (eds.), *Serengeti: Dynamics of an Ecosystem.* Chicago: University of Chicago Press: 310–352; Dublin, H. T. (1991) "Dynamics of the Serengeti-Mara woodlands: an historical perspective." *Forest and Conservation History* 35: 169–178; Sinclair, A. R. E. (1995) "Equilibria in plant-herbivore interactions." In A. R. E. Sinclair and P. Arcese (eds.), *Serengeti II: Dynamics, Management, and Conservation of an Ecosystem.* Chicago: University of Chicago Press: 91–113.

19. The influence of wildebeest numbers on the area burned is shown in Sinclair, A. R. E., J. G. C. Hopcraft, H. Olff, S. A. R. Mduma, K. A. Galvin, and G. J. Sharam. (2008) "Historical and future changes to the Serengeti ecosystem." In A. R. E. Sinclair, C. Packer, S. A. R. Mduma, and J. M. Fryxell (eds.), *Serengeti III: Human Impacts on Ecosystem Dynamics.* Chicago: University of Chicago Press: 7–46.

20. Increase in tree numbers is given in note 13 above and Packer, C., R. Hilborn, A. Mosser, et al. (2005) "Ecological change, group territoriality and population dynamics in Serengeti lions." *Science* 307: 390–393.

21. See note 13. Locations where lions kill prey are described in Hopcraft, J. G. C., A. R. E. Sinclair, and C. Packer. (2005) "Prey accessibility outweighs prey abundance for the location of hunts in Serengeti lions." *Journal of Animal Ecology* 74: 559–566.

22. See Holdo, R. M., R. D. Holt, M. B. Coughenour, and M. E. Ritchie. (2007) "Plant productivity and soil nitrogen as a function of grazing, migration and fire in an African savanna." *Journal of Ecology* 95: 115–128; Anderson, T. M., M. E. Ritchie, E. Mayemba, S. Eby, J. B. Grace, and S. J. McNaughton. (2007) "Forage nutritive quality in the Serengeti ecosystem: the roles of fire and herbivory." *American Naturalist* 170: 343–357.

23. Regeneration of trees is demonstrated in Dublin, H. T. (1991) "Dynamics of the Serengeti-Mara woodlands: an historical perspective." *Forest and Conservation History* 35: 169–178; Sharam, G., A. R. E. Sinclair, and R. Turkington. (2006) "Establishment of broad-leaved thickets in Serengeti, Tanzania: The influence of fire, browsers, grass competition, and elephants." *Biotropica* 38: 599–605; Sharam, G. J., A. R. E. Sinclair, R. Turkington, and A. L. Jacob. (2009) "The savanna tree *Acacia polyacantha* facilitates the establishment of riparian forests in Serengeti National Park, Tanzania." *Journal of Tropical Ecology* 25: 31–40.

24. For effects of fire and grazing on bird communities, see Nkwabi, A. K., A. R. E. Sinclair, K. L. Metzger, and S. A. R. Mduma. (2011) "Disturbance, ecosystem function and compensation: the influence of wildfire and grazing on the avian community in the Serengeti ecosystem, Tanzania." *Austral Ecology* 36: 403–412. In Australia, see Woinarski, J. C. Z. (1990) "Effects of fire on the bird communities of tropical woodlands and open forests in northern Australia." *Austral Ecology* 15: 1–22; Wooller, R. D., and K. S. Brooker. (1980) "The effects of controlled burning on some birds of the understorey in Karri forests." *Emu* 80: 165–166.

25. Gottschalk, T. (2002) "Birds of Grumeti river forest in Serengeti National Park, Tanzania." *African Bird Club Bulletin* 9: 101–104.

26. See Sinclair, A.R.E. (1977) *The African Buffalo. A Study of Resource Limitation of Populations.* Chicago: University of Chicago Press; Eby, S. L. (2010) "Fire and the Reasons for Its Influence on Mammalian Herbivore Distributions in an African Savanna Ecosystem." PhD thesis, Syracuse: Syracuse University.

27. For effects of green flush on the behavior of ungulates, see Mduma, S. A. R., and A. R. E. Sinclair. (1994) "The function of habitat selection by oribi in Serengeti, Tanzania." *African Journal of Ecology* 32: 16–29; Wilsey, B. J. (1996) "Variation in use of green flushes following burns among African ungulate species: the importance of body size." *African Journal of Ecology* 34: 32–38.

28. See notes 26 and 27.

29. See note 13.

30. See note 24.

Chapter 9: Continuous Change in Ecosystems

1. Leakey, M. D., and R. L. Hay. (1979) "Pliocene Footprints in the Laetoli Beds at Laetoli, Northern Tanzania." *Nature* 278: 317–323.

2. Sinclair, A.R.E., M. D. Leakey, and M. Norton-Griffiths. (1986) "Migration and hominid bipedalism." *Nature* 324: 307–308.

3. Changes in the ecology of ancient Serengeti are published in Peters, C. R., R. J. Blumenschine, R. L. Hay, et al. (2008) "Paleoecology of the Serengeti-Mara ecosystem." In A. R. E. Sinclair, C. Packer, S. A. R. Mduma, and J. M. Fryxell (eds.), *Serengeti III: Human Impacts on Ecosystem Dynamics.* Chicago: University of Chicago Press: 47–94.

4. Cohen, A. S., J. R. Stone, K. R. M. Beuning, et al. (2007) "Ecological consequences of early Late Pleistocene megadroughts in tropical Africa." *Proceedings of the National Academy of Sciences, USA* 104: 16422–16427.

5. See note 4.

6. See note 4.

7. Johnson, T. C., C. A. R. Scholz, M. R. et al. (1996) "Late Pleistocene desiccation of Lake Victoria and rapid evolution of cichlid fishes." *Science* 273: 1091–1093.

8. Olago, D. O. (2001) "Vegetation changes over palaeo-time scales in Africa." *Climate Research* 17: 105–121.

9. Bonnefille, R., and F. Chalié. (2000) "Pollen-inferred precipitation time-series from equatorial mountains, Africa, the last 40 kyr BP." *Global and Planetary Change* 26: 25–50. See also Bonnefille, R., and U. Mohammed. (1994) "Pollen-inferred climatic fluctuations in Ethiopia during the last 3000 yrs." *Palaeogeography, Palaeoclimatology and Palaeoecology* 109: 331–343; Stager, J. C., B. Cumming, and L. Meeker. (1997) "A high-resolution 11,400-yr diatom record from Lake Victoria, east Africa." *Quaternary Research* 47: 81–89.

10. From 12–11 kya (kya = thousand years ago), there were warmer and wetter conditions with advancement of lowland forest. A cold, dry episode 11–10 kya was followed by a sustained wet period to about 5 kya, termed the African Humid Period, when rainfall increased to possibly double that of today and lakes rose some 100m above present levels. Lake Chad, in West Africa, for example, was 25 times greater in area and about the same size as the present Caspian Sea. Thompson, L. G., E. Mosley-Thompson, M. E. Davis, et al. (2002) "Kilimanjaro Ice Core Records: Evidence of Holocene Climate Change in Tropical Africa." *Science* 298: 589–593.

11. During the postglacial humid period there were three sharp shifts to drought conditions at 8.3, 5.2, and 4.0 kya. The one at 5.2 kya, which lasted 100 years, coincided with the advance of the Sahara Desert and the rise of human societies in the Nile valley. The drought at 4.0 kya was severe and lasted some 300 years. It caused the collapse of the Egyptian Old Kingdom. See Cullen, H. M., P. B. deMenocal, S. Hemming, G. Hemming, F. H. Brown, T. Guilderson, and F. Sirocko. (2000) "Climate change and the collapse of the Akkadian empire: Evidence from the deep sea." *Geology* 28: 379–382.

12. Gillespie, R., F. A. Street-Perrott, and R. Switsur. (1983) "Post-glacial arid episodes in Ethiopia have implications for climate prediction." *Nature* 306: 680–683. See also note 8.

13. For the Medieval Warm Period, see Tyson, P. D., and J. A. Lindesay. (1992) "The climate of the last 2000 years in southern Africa." *Holocene* 2: 271–278.

14. For the Little Ice Age, see Bradley, R. S., and P. Jones. (1993) "'Little Ice Age' summer temperature variations: their nature and relevance to recent global warming trends." *Holocene* 3: 367–376.

15. Water levels at Lake Naivasha, Kenya, were high when temperatures were lower during the three solar minima, named the Wolf, Spörer, and Maunder minima. The earliest (Wolf) period showed high levels from 1290 to 1370, while the last (Maunder) showed lake overflow during the 1670–1780 period that was interrupted by three prolonged dry episodes. These were 1380–1420, 1560–1620, and 1760–1840. Over the past 200 years, we see that the climate was drier in the early 1800s than subsequently but then became very wet in 1890–1910, and above average in 1885–1940 compared with present (2020) times. See Riehl, H., and J. Meitin. (1979) "Discharge of the Nile River: A Barometer of Short-Period Climate Variation." *Science* 206: 1178–1179. See also Verschuren, D., K. R. Laird, and B. F. Cumming. (2000) "Rainfall and drought in equatorial East Africa during the past 1,100 years." *Nature* 403: 410–414; Verschuren, D. (2001) "Reconstructing fluctuations of a shallow East African lake during the past 1800 years from sediment stratigraphy in a submerged crater basin." *Journal of Paleolimnology* 25: 297–311.

16. Sinclair, A. R. E., J. G. C. Hopcraft, H. Olff, S. A. R. Mduma, K. A. Galvin, and G. J. Sharam. (2008) "Historical and future changes to the Serengeti ecosystem." In A. R. E. Sinclair, C. Packer, S. A. R. Mduma, and J. M. Fryxell (eds.), *Serengeti III: Human Impacts on Ecosystem Dynamics*. Chicago: University of Chicago Press: 7–46.

17. See Hulme, M., R. Doherty, T. Ngara, M. New, and D. Lister. (2001) "Africa climate change: 1900–2100." *Climate Research* 17: 145–168; Nicholson, S. E., D. J. Nash, B. M. Chase, et al. (2013) "Temperature variability over Africa during the last 2000 years." *Holocene* 23: 1085–1094.

18. Sinclair. A. R. E., K. L. Metzger, J. M. Fryxell, et al. (2013) "Asynchronous food web pathways buffer the response of Serengeti predators to El Niño Southern Oscillation." *Ecology* 94: 1123–1130. See also Ritchie, M. E. (2008) "Global environmental changes and their impact on the Serengeti." In A. R. E. Sinclair, C. Packer, S. A. R. Mduma, and J. M. Fryxell (eds.), *Serengeti III: Human Impacts on Ecosystem Dynamics*. Chicago: University of Chicago Press: 183–208.

19. Davis, M., and R. G. Shaw. (2001) "Range shifts and adaptive responses to Quaternary climate change." *Science* 292: 673–679.

20. Gillson, L. (2004) "Testing non-equilibrium theories in savannas: 1400 years of vegetation change in Tsavo National Park, Kenya." *Ecological Complexity* 1: 281–298.

21. S. E. White stated in 1913 that tribes were limited by tsetse fly west of the Ikorongo hills. White, S. E. (1915) *The Rediscovered Country*. New York: Doubleday, Page and Co. Movements of peoples are described in Verschuren, D., K. R. Laird, and B. F. Cumming. (2000) in note 13 above.

22. The arrival of Maasai in the 1850s is recorded in Grant, H. St. J. (1957) *A Report on Human Habitation in the Serengeti National Park*. Dar es Salaam, Tanganyika: Government Printer. I refer to this in Sinclair, A. R. E. (2012) *Serengeti Story*. Oxford: Oxford University Press.

23. The impacts of fire on regeneration of Mara River forests are reported in Dublin, H. T. (1991) "Dynamics of the Serengeti-Mara woodlands: an historical perspective." *Forest and Conservation History* 35: 169–178; Sharam, G., A. R. E. Sinclair, and R. Turkington. (2006) "Establishment of broad-leaved thickets in Serengeti, Tanzania: The influence of fire, browsers, grass competition, and elephants." *Biotropica* 38: 599–605.

24. See Sharam, G. J., A. R. E. Sinclair, R. Turkington, and A. L. Jacob. (2009) "The savanna tree *Acacia polyacantha* facilitates the establishment of riparian forests in Serengeti National Park, Tanzania." *Journal of Tropical Ecology* 25: 31–40.

25. The consensus across Africa in the mid-1960s was that elephants were destroying the woodlands of national parks, so culling was justified. For Murchison Falls National Park, Uganda: Laws, R. M., I.S.C. Parker, and R.C.B. Johnstone. (1975) *Elephants and Their Habitats*. London: Clarendon Press. For Kruger National Park, South Africa: Pienaar, U. de V., P. W. van Wyk, and N. Fairall. (1966) "An Aerial Census of Elephant and Buffalo in the Kruger National Park and the Implications Thereof on Intended Management Schemes." *Koedoe* 9: 40–108.

26. The case for culling in Tsavo National Park, Kenya, was presented by Laws, R. M. (1969) "The Tsavo Research Project." *Journal of Reproduction and Fertility* Supplement 6: 495–531. The counterargument is evident in the data of Corfield, T. F. (1973) "Elephant Mortality in Tsavo National Park, Kenya." *East African Wildlife Journal* 11: 339–368.

27. Culling in the Serengeti was being mooted in the 1960s but was never accomplished; see Lamprey, H. F., P. E. Glover, M. I. M. Turner, and R. H. V. Bell. (1967) "Invasion of the Serengeti National Park by elephants." *East African Wildlife Journal* 5: 151–166; Watson, R. M., and R. H. V. Bell. (1969) "The distribution, abundance and status of elephant in the Serengeti region of northern Tanzania." *Journal of Applied Ecology* 6: 115–132.

28. The first study of fire is Norton-Griffiths, M. (1979) "The influence of grazing, browsing, and fire on the vegetation dynamics of the Serengeti." In A. R. E. Sinclair and M. Norton-Griffiths (eds.), *Serengeti: Dynamics of an Ecosystem.* Chicago: University of Chicago Press: 310–352.

29. The photos by Syd Downey in 1944 and ours in 1984 can be seen in Dublin, H. T. (1991) "Dynamics of the Serengeti-Mara woodlands: an historical perspective," *Forest and Conservation History* 35: 169–178.

30. Martin Johnson and Osa Johnson set out on a series of photographic adventures to New Guinea, Borneo, and Africa during the years 1922–37. They recorded their exploits in a number of books. On 12 January 1937 Martin Johnson was killed in an airplane crash at Burbank, California, but Osa survived. She took no further expeditions but wrote her memoirs. She died in 1953. Their joint biography can be found in Stott, K. W. (1978) *Exploring with Martin and Osa Johnson.* Chanute, KS: Martin and Osa Johnson Safari Museum Press. Stott himself had met both the Johnsons and came to know Osa well. I talked with him in 1981 about the Serengeti expeditions. See Johnson, M. (1929) *Lion: African Adventure with the King of Beasts.* New York: G. P. Putnam's Sons.

31. Baumann, O. (1894) *Durch Massailand zur Nilquelle.* Berlin. Reprinted 1968. New York: Johnson Reprint Corporation.

32. The expeditions are reported in the books by Martin and Osa Johnson. Johnson, M. (1928) *Safari.* New York: G. P. Putnam's Sons; Johnson, M. (1929) *Lion: African Adventure with the King of Beasts.* New York: G. P. Putnam's Sons; Johnson, M. (1935) *Over African Jungles.* New York: Harcourt, Brace.

33. The consequences of rinderpest on human and cattle populations, including the spread of ancillary diseases, is beautifully documented in Ford, J. (1971) *The Role of Trypanosomiases in African Ecology.* Oxford: Clarendon Press.

34. Huxley, J. (1936) *Africa View.* London: Chatto and Windus. Julian Huxley toured the Shinyanga and Mwanza areas in 1929 to make a report to the British government. He comments on tsetse-fly control, including burning, on pages 75–85.

35. For the relation between wildebeest grazing and area burned, see Sinclair, A. R. E., J. G. C. Hopcraft, H. Olff, S. A. R. Mduma, K. A. Galvin, and G. J. Sharam. (2008) "Historical and future changes to the Serengeti ecosystem." In A. R. E. Sinclair, C. Packer, S. A. R. Mduma, and J. M. Fryxell (eds.), *Serengeti III: Human Impacts on Ecosystem Dynamics.* Chicago: University of Chicago Press: 7–46.

36. Changes in trees since the 1920s are given in Sinclair, A.R.E. (1995) "Equilibria in plant-herbivore interactions." In A. R. E. Sinclair and P. Arcese (eds.), *Serengeti II: Dynamics, Management, and Conservation of an Ecosystem.* Chicago: University of Chicago Press: 91–113.

37. See note 31.

38. White, S. E. (1915) *The Rediscovered Country.* New York: Doubleday, Page.

39. The traders found Wandorobo elephant hunters as they traveled across Serengeti in the 1870s. See Wakefield, T. (1870) "Routes of native caravans from the coast to the interior of East

Africa." *Journal of the Royal Geographical Society* 11: 303–338; Wakefield, T. (1882) "New routes through Masai country." *Proceedings of the Royal Geographical Society* 4: 742–747; Farler, J. P. (1882) "Native routes in East Africa from Pangani to the Masai country and the Victoria Nyanza." *Proceedings of the Royal Geographical Society* 4: 730–742. Also reported in Fosbrooke, H. A. (1968) "Elephants in the Serengeti National Park: An Early Record." *East African Wildlife Journal* 6: 150–152.

40. See Shaw, P., A. R. E. Sinclair, K. Metzger, A. Nkwabi, S. A. R. Mduma, and N. Baker. (2010) "Range expansion of the globally vulnerable Karamoja apalis *Apalis karamojae* in the Serengeti ecosystem." *African Journal of Ecology* 48: 751–758.

41. Beale, C. M., N. E. Baker, M. J. Brewer, and J. J. Lennon. (2013) "Protected area networks and savannah bird biodiversity in the face of climate change and land degradation." *Ecology Letters* 16: 1061–1068.

Chapter 10: Appearance of Multiple States and Rapid Shifts in Ecosystems

1. Syd Downey saw the Mara Triangle as his personal hunting area. See Herne, B. (1999) *White Hunters: The Golden Age of African Safaris.* New York: Henry Holt and Co., pages 154–157.

2. The photos taken by Syd Downey in 1944 and ours in 1984 can be seen in Dublin, H. T. (1991) "Dynamics of the Serengeti-Mara woodlands: an historical perspective." *Forest and Conservation History* 35: 169–178.

3. For Dublin's work on Mara tree dynamics, see Dublin, H. T. (1995) "Vegetation dynamics in the Serengeti-Mara ecosystem: the role of elephants, fire and other factors." In A. R. E. Sinclair and P. Arcese (eds.), *Serengeti II: Dynamics, Management, and Conservation of an Ecosystem.* Chicago: University of Chicago Press: 71–90.

4. For changes in the elephant population, see Sinclair, A. R. E., J. G. C. Hopcraft, H. Olff, S. A. R. Mduma, K. A. Galvin, and G. J. Sharam. (2008) "Historical and future changes to the Serengeti ecosystem." In A. R. E. Sinclair, C. Packer, S. A. R. Mduma, and J. M. Fryxell (eds.), *Serengeti III: Human Impacts on Ecosystem Dynamics.* Chicago: University of Chicago Press: 7–46.

5. The collapse of the economy as a result of political ideology led to the massive increase in ivory and meat poaching in the 1980s. See chapters 8 and 9 in Sinclair, A. R. E. (2012) *Serengeti Story.* Oxford: Oxford University Press. For the numbers of elephant and rhino killed in Serengeti, see Sinclair, A. R. E. (1995) "Serengeti past and present." In A.R.E. Sinclair and P. Arcese (eds.), *Serengeti II: Dynamics, Management, and Conservation of an Ecosystem.* Chicago: University of Chicago Press: 3–30.

6. For multiple states involving elephants and trees, see Dublin, H. T., A. R. E. Sinclair, and J. McGlade. (1990) "Elephants and fire as causes of multiple stable states for Serengeti-Mara woodlands." *Journal of Animal Ecology* 59: 1157–1164.

7. The theory of multiple states is outlined in Holling, C. S. (1973) "Resilience and stability of ecological systems." *Annual Review of Ecology & Systematics* 4:1–23; Sinclair, A. R. E. (1989) "Population regulation in animals." In J. M. Cherrett (ed.), *Ecological Concepts.* British Ecological Society Symposium 29: 197–241; Knowlton, N. (1992) "Thresholds and multiple stable states

in coral reef community dynamics." *American Zoologist* 32: 674–682; Beisner, B. E., D. T. Haydon, and K. Cuddington. (2003) "Alternative stable states in ecology." *Frontiers in Ecology and Environment* 1: 376–382.

8. Holling, C. S. (1973) in note 7 above.

9. For modeling of the wildebeest migrant population, see Fryxell, J. M., J. Greever, and A. R. E. Sinclair. (1988) "Why are migratory ungulates so abundant?" *American Naturalist* 131: 781–798.

10. Allan Brooks, a Canadian from British Columbia, was the first scientist in Serengeti, posted by the British Colonial Office at the request of E. B. Worthington in 1951. Worthington originally worked on the East African lakes in the 1930s, then with the Food and Agriculture Organization of the United Nations in Rome in the 1950s, and later organized the International Biological Program in the 1960s. In Arusha the game warden assigned Brooks to looking at Thomson's gazelle in Serengeti since this was the dominant species at that time. Together with Worthington and his family, and helped by the warden, John Blower, he conducted two ground censuses across the plains, in September 1952 and February 1953. The first, in the dry season, produced nothing, but the second produced results that could be placed on modern maps and included population numbers computed using standard statistical techniques. Brooks retired to Saltspring Island, British Columbia, in the 1970s and died in 2000. His widow gave us the raw data in 2013, which is now with the Beaty Biodiversity Museum, UBC. See Brooks, A. C. (1961) *A Study of the Thomson's Gazelle* (Gazella thomsonii *Gunther) in Tanganyika*. London: Colonial Research Publications No. 25, H.M.S.O.

11. Work on forest dynamics by Greg Sharam includes Sharam, G., A. R. E. Sinclair, and R. Turkington. (2006) "Establishment of broad-leaved thickets in Serengeti, Tanzania: The influence of fire, browsers, grass competition, and elephants." *Biotropica* 38: 599–605; Sharam, G, A. R. E. Sinclair, and R. Turkington. (2009) "Serengeti birds maintain forests by inhibiting seed predators." *Science* 325: 51; Turkington, R., G. Sharam, and A. R. E. Sinclair. (2015) "Biodiversity and the dynamics of riverine forests in Serengeti." In A. R. E. Sinclair, K. L. Metzger, J. M. Fryxell, and S. A. R. Mduma (eds.), *Serengeti IV: Sustaining Biodiversity in a Coupled Human-Natural System*. Chicago: University of Chicago Press: 235–264.

12. For birds as influencers of forest structure, see Clout, M. N., and J. J. Hay. (1989) "The importance of birds as browsers, pollinators and seed dispersers in NewZealand forests." *New Zealand Journal of Ecology* 12, supplement: s27–33; Sun, C., A. R. Ives, H. J. Kraeuter, and T. C. Moermond. (1997) "Effectiveness of three turacos as seed dispersers in a tropical montane forest." *Oecologia* 112: 94–103.

13. The view that Africa was in a pristine state when European colonists first arrived at the turn of the twentieth century is well illustrated in Noel Simon's 1962 *Between the Sunlight and the Thunder*. London: Collins. See also Laws, R. M. (1969) "The Tsavo Research Project." *Journal of Reproduction and Fertility* Supplement 6: 495–531; and Laws, R. M., I. S. C. Parker, and R. C. B. Johnstone. (1975) *Elephants and their Habitats* London: Clarendon Press. The opinion that the pristine state of savanna in Serengeti was a parkland of beautiful, large trees being destroyed by elephants that never should have been there is reported in Croze, H., A. K. K. Hillman, and E. M. Lang (1981) "Elephants and their habitats: How do they tolerate each other?" In C. W. Fowler and T. D. Smith (eds.), *Dynamics of Large Animal Populations* New York: Wiley: 297–316.

14. Multiple states in different ecosystems can be seen in Augustine, D. J., L. E. Frelich, and P. A. Jordan. (1998) "Evidence for two alternate stable states in an ungulate grazing system." *Ecological Applications* 8: 1260–1269. A more theoretical overview of multiple stable states can be found in Beisner, B. E., et al. (2003) in note 7 above. See also a comprehensive theoretical treatment and real-world examples in Petraitis, P. S. (2013) *Multiple Stable States in Natural Ecosystems.* Oxford: Oxford University Press.

Chapter 11: The Fundamental Principle of Regulation, and Future Directions

1. More biodiversity within and among trophic levels resulted in higher regulation (and lower variability) of individual species populations, or a greater variability between species population sizes but a constancy of the total numbers over all species (due to compensation). This is outlined in note 15 of chapter 7.

2. Olff, H., and J. G. C. Hopcraft. (2008) "The resource basis of human-wildlife interaction." In A. R. E. Sinclair, C. Packer, S. A. R. Mduma, and J. M. Fryxell (eds.), *Serengeti III: Human Impacts on Ecosystem Dynamics.* Chicago: University of Chicago Press: 95–134; Metzger, K. L., A. R. E. Sinclair, A. Macfarlane, M. B. Coughenour, and J. Ding. (2015) "Scales of change in the greater Serengeti ecosystem." In A. R. E. Sinclair, K. L. Metzger, J. M. Fryxell, and S. A. R. Mduma (eds.), *Serengeti IV: Sustaining Biodiversity in a Coupled Human-Natural System.* Chicago: University of Chicago Press: 33–72.

3. Verchot, L. V., N. L. Ward, J. Belnap, et al. (2015) "From bacteria to elephants: Effects of land-use legacies on biodiversity and ecosystem processes in the Serengeti-Mara ecosystem." In A. R. E. Sinclair, K. L. Metzger, J. M. Fryxell, and S. A. R. Mduma (eds.), *Serengeti IV: Sustaining Biodiversity in a Coupled Human-Natural System.* Chicago: University of Chicago Press: 195–234.

4. Veldhuis, M. P., M. E. Ritchie, J. O. Ogutu, et al. (2019) "Cross-boundary human impacts compromise the Serengeti-Mara ecosystem." *Science* 363: 1424–1428.

5. Larsen, F., J. G. C. Hopcraft, N. Hanley, et al. (2020) "Wildebeest migration drives tourism demand in the Serengeti." *Biological Conservation* 248. https://doi.org/10.1016/j.biocon.2020 .108688.

6. Morrison, T. A., R. M. Holdo, D. M. Rugemalila, M. Nzunda, and T. M. Anderson. (2018) "Grass competition overwhelms effects of herbivores and precipitation on seedling establishment in Serengeti." *Journal of Ecology* 107: 216–228.

7. Anderson, T. M., T. Morrison, D. Rugemalila, and R. Holdo. (2015) "Compositional decoupling of savanna canopy and understory tree communities in Serengeti." *Journal of Vegetation Science* 26: 385–394.

8. Herlocker, D. J. (1976) "Structure, Composition, and Environment of Some Woodland Vegetation Types of the Serengeti National Park, Tanzania." PhD dissertation, College Station: Texas A&M University.

9. Morrison, T.A.R., R. Holdo, and T. M. Anderson. (2016) "Elephant damage, not fire or rainfall, explains mortality of overstory trees in Serengeti." *Journal of Ecology* 104: 409–418; Rugemalila, D., R. Holdo, and T. M. Anderson. (2016) "Precipitation and elephants, not fire, shape tree community composition in Serengeti. *Biotropica* 48: 476–482.

10. Eby, S., J. Demperwolf, R. M. Holdo, and K. L. Metzger. (2015) "Fire in the Serengeti ecosystem: History, drivers, and consequences." In A. R. E. Sinclair, K. L. Metzger, J. M. Fryxell, and S. A. R. Mduma (eds.), *Serengeti IV: Sustaining Biodiversity in a Coupled Human-Natural System*. Chicago: University of Chicago Press: 33–72.

11. Anderson, T. M., J. G. C. Hopcraft, S. L. Eby, M. E. Ritchie, J. B. Grace, and H. Olff. (2010) "Landscape-scale analyses suggest both nutrient and antipredator advantages to Serengeti herbivore hotspots." *Ecology* 91:1519–1529; Hopcraft, J. G. C., T. M. Anderson, S. Pérez-Vila, E. Mayemba, and H. Olff. (2012) "Body size and the division of niche space: food and predation differentially shape the distribution of Serengeti grazers." *Journal of Animal Ecology* 81: 201–213.

12. Swanson, A., M. Kosmala, C. Lintott, R. Simpson, A. Smith, and C. Packer. (2015) "Snapshot Serengeti, high-frequency, fine-scale annotated camera trap images of 40 mammalian species in an African savanna." *Scientific Data* 2: 150026. doi:10.1038/sdata.2015.26; Swanson, A., J. Forester, T. Arnold, M. Kosmala, and C. Packer. (2016) "In the absence of a landscape of fear: how lions, hyenas, and cheetahs coexist." *Ecology and Evolution* 6: 8534–8545. doi:10.1002/ece3.2569; Swanson, A., M. Kosmala, C. Lintott, C. Packer. (2016) "A generalized approach for producing, quantifying, and validating citizen science data from wildlife images." *Conservation Biology* 30: 520–531. doi:10.1111/cobi.12695; Hepler, S. A., R. Erhardt, and T. M. Anderson. (2018) "Identifying drivers of spatial variation in occupancy with limited replication camera trap data." *Ecology* 99: 2152–2158; Beaudrot, L., M. S. Palmer, T. M. Anderson, and C. Packer. (2020) "Mixed-species groups of Serengeti grazers: a test of the stress gradient hypothesis." *Ecology* 101. https://doi.org/10.1002/ecy.3163.

13. Swanson, A., T. Caro, H. Davies-Mostert, et al. (2014) "Cheetahs and wild dogs show contrasting patterns of suppression by lions." *Journal of Animal Ecology* 83: 1418–1427. doi:10.1111/1365-2656.12231.

14. de Visser, S. N., B. P. Freymann, and H. Olff. (2011) "The Serengeti food web: empirical quantification and analysis of topological changes under increasing human impact." *Journal of Animal Ecology* 80: 484–494.

15. Sinclair, A. R. E. (1975) "The resource limitation of trophic levels in tropical grassland ecosystems." *Journal of Animal Ecology* 44: 497–520.

16. Freymann, B. P., R. Buitenwerf, O. DeSouza, and H. Olff. (2008) "The importance of termites (*Isoptera*) for the recycling of herbivore dung in tropical ecosystems: a review." *European Journal of Entomology* 105: 165–173; Freymann, B. P., S. N. de Visser, and H. Olff. (2010) "Spatial and temporal hotspots of termite-driven decomposition in the Serengeti." *Ecography* 33: 443–450.

17. Nkwabi, A. K., A. R. E. Sinclair, K. L. Metzger, and S. A. R. Mduma. (2011) "Disturbance, ecosystem function and compensation: the influence of wildfire and grazing on the avian community in the Serengeti ecosystem, Tanzania." *Austral Ecology* 36: 403–412; Nkwabi, A. K., A. R. E. Sinclair, and S. A. R. Mduma. (2015) "The effect of natural disturbances on the avian community of the Serengeti woodlands." In A. R. E. Sinclair, K. L. Metzger, J. M. Fryxell, and S. A. R. Mduma (eds.), *Serengeti IV: Sustaining Biodiversity in a Coupled Human-Natural System*. Chicago: University of Chicago Press: 395–418.

18. See Vanderlaan, A. S. M., and C. T. Taggart. (2007) "Vessel collisions with whales: The probability of lethal injury based on vessel speed." *Marine Mammal Science* 23: 144–156.

19. Holdo, R. M., A. R. E. Sinclair, K. L. Metzger, et al. (2009) "A Disease-Mediated Trophic Cascade in the Serengeti and its Implications for Ecosystem C." *PLoS Biology* 7: e1000210.

20. Sinclair, A. R. E. (1977) *The African Buffalo: A Study of Resource Limitation of Populations.* Chicago: University of Chicago Press.

21. See note 20.

22. See note 12.

Chapter 12: Threats to the Serengeti

1. Tingvold et al. have shown that elephants have higher stress hormones not only when they leave the park but even when they approach the boundaries of the park prior to leaving. See Tingvold, H. G., R. Fyumagwa, C. Bech, L. F. Baardsen, H. Rosenlund, and E. Røskaft. (2013) "Determining adrenocortical activity as a measure of stress in African elephants (*Loxodonta africana*) in relation to human activities in Serengeti ecosystem." *African Journal of Ecology* 51: 580–589. doi:10.1111/aje.12069. Female elephants in areas with a high risk of poaching or that belong to a group where the social structure has been damaged by poaching have decreased reproductive outputs. See Gobush, K. S., B. M. Mutayoba, and S. K. Wasser. (2008) "Long-term impacts of poaching on relatedness, stress physiology, and reproductive output of adult female African elephants." *Conservation Biology* 22: 1590–1599.

2. The flow of water is reported by Eric Wolanski and Emmanuel Gereta. See Mnaya, B., Y. Kiwango, E. Gereta, and E. Wolanski. (2011) "Ecohydrology-based planning as a solution to address an emerging water crisis in the Serengeti ecosystem and Lake Victoria." In H. S. Elliot and L. E. Martin (eds.), *River Ecosystems: Dynamics, Management and Conservation.* NOVA Science publishers: 233–258. See also Mnaya, B., M. Mtahiko, and E. Wolanski. (2017) "The Serengeti will die if Kenya dams the Mara River." *Oryx* 5: 581–583. doi:10.1017/S0030605317001338.

3. The widening of the Mara River is documented in Sinclair, A. R. E., J. G. C. Hopcraft, H. Olff, S. A. R. Mduma, K. A. Galvin, and G. J. Sharam. (2008) "Historical and future changes to the Serengeti ecosystem." In A. R. E. Sinclair, C. Packer, S. A. R. Mduma, and J. M. Fryxell (eds.), *Serengeti III: Human Impacts on Ecosystem Dynamics.* Chicago: University of Chicago Press: 7–46.

4. Loss of riverine forest is reported in Turkington, R., G. Sharam, and A. R. E. Sinclair. (2015) "Biodiversity and the dynamics of riverine forests in Serengeti." In A. R. E. Sinclair, K. L. Metzger, J. M. Fryxell, and S. A. R. Mduma (eds.), *Serengeti IV: Sustaining Biodiversity in a Coupled Human-Natural System.* Chicago: University of Chicago Press: 235–264.

5. The story of the collapse of the Aral Sea is reported by Dene-Hern Chen of *National Geographic* magazine on 16 March 2018. See https://news.nationalgeographic.com/2018/03 /north-aral-sea-restoration-fish-kazakhstan.

6. The impacts of agriculture, fences, and pastoralism on the borders of and within Serengeti are presented in Veldhuis, M. P., M. E. Ritchie, J. O. Ogutu, et al. (2019) "Cross-boundary human impacts compromise the Serengeti-Mara ecosystem." *Science* 363: 1424–1428.

7. Alfred Kikoti placed a collar on a male elephant on the slopes of Mt. Kilimanjaro and followed it some 600 km west across the Rift Valley and into Serengeti, where it remained in 2014.

8. Wild dog movements are being recorded by Ernest Eblate and Emmanuel Masenga. Simon Mduma related the long-distance movements in January 2016.

9. Packer, C., A. Loveridge, S. Canney, et al. (2013) "Conserving large carnivores: dollars and fence." *Ecology Letters* 16: 635–641; Durant, S. M., M. S. Becker, S. Bashir, et al. (2015) "Developing fencing policies for dry land ecosystems." *Journal of Applied Ecology* 52: 544–551. doi: 10.1111/1365-2664.12415.

10. For ecological baselines, see Sinclair A. R. E., R. P. Pech, J. M. Fryxell, et al. (2018) "Predicting and assessing progress in the restoration of ecosystems." *Conservation Letters* 11: e12390; Arcese, P., and A. R. E. Sinclair. (1997) "The role of protected areas as ecological baselines." *Journal of Wildlife Management* 61: 587–602.

11. For the decline of Mara Reserve ungulates, see Ogutu, J. O., H.-P. Piepho, H. T. Dublin, N. Bhola, and R. S. Reid. (2009) "Dynamics of Mara–Serengeti ungulates in relation to land use changes." *Journal of Zoology* 278: 1–14; Ogutu, J. O., N. Owen-Smith, H. P. Piepho, and M. Y. Said. (2011) "Continuing wildlife population declines and range contraction in the Mara region of Kenya during 1977–2009." *Journal of Zoology* 285: 99–109; Green, D. S., E. F. Zipkin, D. C. Incorvaia, and K. E. Holekamp. (2019) "Long-term ecological changes influence herbivore diversity and abundance inside a protected area in the Mara-Serengeti ecosystem." *Global Ecology and Conservation* 20: (e00697).

12. Larsen, F., J. G. C. Hopcraft, N. Hanley, et al. (2020) "Wildebeest migration drives tourism demand in the Serengeti." *Biological Conservation* 248. https://doi.org/10.1016/j.biocon.2020.108688.

13. Changes in human populations around the boundaries of the Serengeti ecosystem are reported in Campbell, K., and H. Hofer. (1995) "People and wildlife spatial dynamics and zones of interaction." In A. R. E. Sinclair and P. Arcese (eds.), *Serengeti II: Dynamics, Management, and Conservation of an Ecosystem.* Chicago: University of Chicago Press: 534–570; and Estes, A. B., T. Kuemmerle, H. Kushnir, V. C. Radeloff, and H. H. Shugart (2015) "Agricultural expansion and human population trends in the Greater Serengeti Ecosystem from 1984 to 2003." In A. R. E. Sinclair, K. L. Metzger, J. M. Fryxell, and S. A. R. Mduma (eds.), *Serengeti IV: Sustaining Biodiversity in a Coupled Human-Natural System.* Chicago: University of Chicago Press: 513–532.

For illegal hunting off-take, see Rentsch, D., R. Hilborn, E. J. Knapp, K. L. Metzger, and M. Loibooki. (2015) "Bushmeat hunting in the Serengeti ecosystem: An assessment of drivers and impact on migratory and non-migratory wildlife." In A. R. E. Sinclair, K. L. Metzger, J. M. Fryxell, and S. A. R. Mduma (eds.), *Serengeti IV: Sustaining Biodiversity in a Coupled Human-Natural System.* Chicago: University of Chicago Press: 649–678.

14. Lembo T., K. Hampson, D. Haydon, et al. (2008) "Exploring reservoir dynamics: a case study of rabies in the Serengeti ecosystem." *Journal of Applied Ecology* 45(4): 1246–1257. doi: 10.1111/j.1365-2664.2008.01468.x; Lembo T., K. Hampson, M. Kaare, et al. (2010) "The feasibility of eliminating canine rabies in Africa: dispelling doubts with data." *PLoS Neglected Tropical Diseases* 4: e626.

15. Viana, M., S. Cleaveland, J. Matthiopoulos, et al. (2015) "Dynamics of a morbillivirus at the domestic-wildlife interface: canine distemper virus in domestic dogs and lions." *Proceedings of the National Academy of Sciences, USA* 112 (5): 1464–1469. doi:10.1073/pnas.1411623112.

16. Casey-Bryars, M., R. Reeve, U. Bastola, et al. (2018) "Waves of endemic foot-and-mouth disease in eastern Africa suggest feasibility of proactive vaccination approaches." *Nature Ecology & Evolution* 2: 1449–1457. doi:10.1038/s41559-018-0636-x.

17. See note 6.

18. Harris, G., S. Thirgood, J. G. C. Hopcraft, J. P. G. M. Cromsigt, and J. Berger. (2009) "Global decline in aggregated migrations of large terrestrial mammals." *Endangered Species Research* 7: 55–76.

19. The collapse of the human-cattle migrations in the Sahel is described in Sinclair, A. R. E., and J. M. Fryxell. (1985) "The Sahel of Africa: ecology of a disaster." *Canadian Journal of Zoology* 63: 987–994.

20. The account of the proposed road across northern Serengeti in 2010 is outlined in Sinclair, A. R. E. (2012) *Serengeti Story*. Oxford: Oxford University Press. For analysis on why the road should not be built, see Dobson, A. P., M. Borner, A. R. E. Sinclair, et al. (2010) "Road will ruin Serengeti." *Nature* 467: 272–274; Sinclair, A. R. E. (2010) "Road proposal threatens existence of Serengeti." *Oryx* 44: 478–479; Holdo, R. M., J. M. Fryxell, A. R. E. Sinclair, A. Dobson, and R. D. Holt. (2011) "Predicted impact of barriers to migration on the Serengeti wildebeest migration. *PLoS ONE* 6(1): e16370. doi:10.1371/journal.pone.0016370.

21. Modern work on the Serengeti migrants is found in Thirgood, S., A. Mosser, M. Borner, et al. (2004) "Can Parks Protect Migratory Ungulates? The Case of the Serengeti Wildebeest." *Animal Conservation* 7: 113–20. The historical distribution of the migrants and other species is presented in Sinclair, A. R. E. (2012) *Serengeti Story*. Oxford: Oxford University Press; and fig. 2.4 in Sinclair, A. R. E., A. Dobson, S. A. R. Mduma, and K. L. Metzger. (2015) "Shaping the Serengeti ecosystem." In A. R. E. Sinclair, K. L. Metzger, J. M. Fryxell, and S. A. R. Mduma (eds.), *Serengeti IV: Sustaining Biodiversity in a Coupled Human-Natural System*. Chicago: University of Chicago Press: 11–29. See also White, S. E. (1915) *The Rediscovered Country*. New York: Doubleday, Page and Co.

22. In Banff National Park, Canada, roadkills involved moose and elk to the extent that a fence was built on either side of the highway. In Hluhluwe-iMfolosi Park, South Africa, a 10-mile road constructed in 2002 has resulted in numerous roadkills of wildlife, especially wild dogs, now a threatened species, but also lions, leopards, buffalo, rhino, and many antelope species. Elephant have been hit and injured. Human fatalities have occurred. The road was constructed with road humps to control vehicle speeds, but cars have been recorded at 120 kph. There are no fences along the whole length, but barriers were placed to reduce the risk of animals crossing in dangerous places. This served to impound animals on the road and actually increased collisions. In general, the park manager, Sihle Nxulalo, has concluded that the road has become a serious problem, as reported by Sue van Rensburg. The effects of the tarmac road constructed in 1972 through Mikumi National Park, Tanzania, show that animals that are disturbed by the road suffer proportionately lower fatalities than species that take no notice of the road. See Newmark, W. D., J. I. Boshe, H. I. Sariko, and G. K. Makumbule. (1996) "Effects of a highway on large mammals in Mikumi National Park, Tanzania." *African Journal of Ecology* 34: 15–31. This scientific observation is relevant to the Serengeti because the huge migrating herds pay little attention to vehicle traffic. Impacts from roads and corridors include habitat loss, intrusion of edge effects in natural areas, isolation of populations, barrier effects, road mortality, and

increased human access. See Benitez-Lopez, A., R. Alkemade, and P. A. Verweij. (2010) "The impacts of roads and other infrastructure on mammal and bird populations: A meta-analysis." *Biological Conservation* 143: 1307–1316.

23. See note 20.

24. The collapse of the Botswana wildebeest migration is reported in Williamson, D., and B. Mbano. (1988) "Wildebeest mortality during 1983 at Lake Xau, Botswana." *African Journal of Ecology* 26: 341–344.

25. The effects of fences in southern Africa can be found in Gadd, M. E. (2012) "Barriers, the beef industry and unnatural selection: a review of the impacts of veterinary fencing on mammals in southern Africa." In M. J. Somers and M. W. Hayward (eds.), *Fencing for Conservation: Restriction of Evolutionary Potential or a Riposte to Threatening Processes?* New York: Springer: 153–186.

26. Other reports on the problems of fences are found in Bartlam-Brooks, H.L.A., M. C. Bonyongo, and S. Harris. (2011) "Will reconnecting ecosystems allow long-distance mammal migrations to resume? A case study of a zebra *Equus burchelli* migration in Botswana." *Oryx* 45: 210–216; Bolger, D. T., W. D. Newmark, T. A. Morrison, and D. F. Doak. (2008) "The need for integrative approaches to understand and conserve migratory ungulates." *Ecology Letters* 11: 63–77. See also Benitez-Lopez, A., et al. (2010) in note 23 and Harris, G., S. et al. (2009) in note 18 above.

27. Infanticide was first documented by Brian Bertram in the early 1970s and later confirmed by the long-term studies of Craig Packer. See Bertram, B. C. R. (1975) "Social factors influencing reproduction in wild lions." *Journal of Zoology* 177: 463–482. See also Packer, C. (2000) "Infanticide is no fallacy." *American Anthropologist* 102: 829–831.

28. For effects of sport hunting on lions within Serengeti National Park, see Borrego, N., A. Ozgul, R. Slotow, and C. Packer. (2018) "Lion population dynamics: do nomadic males matter?" *Behavioral Ecology* 29: 660–666. doi:10.1093/beheco/ary018.

29. For unsustainable hunting of lions outside Serengeti National Park, see Whitman, K., A. M. Starfield, H. S. Quadling, and C. Packer. (2004) "Sustainable trophy hunting of African lions." *Nature* 428: 175–178; Packer, C., H. Brink, B. M. Kissui, H. Maliti, H. Kushnir, and T. Caro. (2010) "Effects of trophy hunting on lion and leopard populations in Tanzania." *Conservation Biology* 25: 142–153.

30. Bird species that have moved into the Serengeti ecosystem from the dry eastern Rift Valley in the past 30 years include the yellow-throated spurfowl, white-bellied go-away-bird, Pangani longclaw, Taita fiscal, and anteater chat. See also Beale, C. M., N. E. Baker, M. J. Brewer, and J. J. Lennon. (2013) "Protected area networks and savannah bird biodiversity in the face of climate change and land degradation." *Ecology Letters* 16: 1061–1068. doi:1111/ele. 12139.

31. For loss of bird species in agriculture, see Sinclair, A. R. E., S. A. R. Mduma, and P. Arcese. (2002) "Protected areas as biodiversity benchmarks for human impacts: agriculture and the Serengeti avifauna." *Proceedings of the Royal Society, London, B.* 269: 2401–2405.

32. Montane-forest species that have declined or disappeared in the Serengeti stretch of the Mara River include the crowned hornbill, Ross's turaco, crested guineafowl, and cinnamon-breasted bee-eater. They are still present in the Kenya portion.

33. See also Pavlacky, D. C., H. P. Possingham, A. J. Lowe, P. J. Prentis, D. J. Green, and A. W. Goldizen. (2012) "Anthropogenic landscape change promotes asymmetric dispersal and limits

regional patch occupancy in a spatially structured bird population." *Journal of Animal Ecology* 81: 940–952.

34. The effect of agriculture on fragile-resilient bird species, and on species at different trophic levels, is reported in Sinclair, A. R. E., A. Nkwabi, S. A. R. Mduma, and F. Magige. (2014) "Responses of the Serengeti avifauna to long-term change in the environment." *Ostrich. Journal of African Ornithology* 85: 1–11.

35. Top-down limitation by birds is demonstrated in Nkwabi, A. K., A. R. E. Sinclair, K. L. Metzger, and S. A. R. Mduma. (2011) "Disturbance, ecosystem function and compensation: the influence of wildfire and grazing on the avian community in the Serengeti ecosystem, Tanzania." *Austral Ecology* 36: 403–412. The removal of top trophic levels can cause outbreaks of rodents. In agriculture, effectively all raptor species, and especially the black-shouldered kite (*Elanus caeruleus*), disappear. At the same time, rodent population density remains six times higher than in native savanna, even during the periods between outbreaks, suggesting that raptors could be controlling rodent numbers between outbreaks. In agriculture, both insects and rodents are pests, a problem for natural-resource management. See Byrom, A. E., M. E. Craft, S. M. Durant, et al. (2014) "Episodic outbreaks of small mammals influence predator community dynamics in an East African savanna ecosystem." *Oikos* 123: 1014–1024; Byrom, A. E., A. J. K. Nkwabi, K. Metzger, et al. (2015) "Anthropogenic stressors influence small mammal communities in tropical East African savanna at multiple spatial scales." *Wildlife Research* 42: 119–131. See also Schmitz, O. J. (2007) "Predator diversity and trophic interactions." *Ecology* 88: 2415–2426; Schneider, F. D., and U. Brose. (2013) "Beyond diversity: how nested predator effects control ecosystem functions." *Journal of Animal Ecology* 82: 64–71.

36. For rapid changes in grassland-savanna habitats of Tsavo National Park, Kenya, see Gillson, L. (2004) "Testing non-equilibrium theories in savannas: 1400 years of vegetation change in Tsavo National Park, Kenya." *Ecological Complexity* 1: 281–298.

Chapter 13: Lessons from Serengeti

1. For downgrading, see Estes, J. A., J. Terborgh, J. S. Brashares, et al. (2011) "Trophic downgrading of planet earth." *Science* 333: 301–306. For upgrading, see Pringle, R. M. (2017) "Upgrading protected areas to conserve wild biodiversity." *Nature* 546: 91–99.

2. The consensus across Africa in the mid-1960s was that elephants were destroying the woodlands of national parks. For Kruger National Park, South Africa: Pienaar, U. de V., P. W. van Wyk, and N. Fairall. (1966) "An Aerial Census of Elephant and Buffalo in the Kruger National Park and the Implications Thereof on Intended Management Schemes." *Koedoe* 9: 40–108; Piennar, U. de V. (1983) "Management by Intervention: The Pragmatic Option." In Owen-Smith, N. R. (ed.). *Management of Large Mammals in African Conservation Areas.* Pretoria: Haum Educational Publishers: 23–26. For Murchison Falls National Park, Uganda: Laws, R. M, I. S. C. Parker, and R. C. B. Johnstone. (1975) *Elephants and Their Habitats.* Oxford: Clarendon Press. For Tsavo National Park, Kenya: Laws, R. M. (1969) "The Tsavo Research Project." *Journal of Reproduction and Fertility* Supplement 6: 495–531.

3. Regulation of Kruger Park elephants is shown in Sinclair, A. R. E., and K. Metzger. (2009) "Advances in wildlife ecology and the influence of Graeme Caughley." *Wildlife Research* 36: 8–15.

4. W. C. Allee was interested in animal behavior, in this context mating behavior. See Allee, W. C. (1938) *The Social Life of Animals*. New York: W. W. Norton & Co.

5. The Allee effect has now been applied to all causes of inverse density dependent regulation in very small populations. See Courchamp, F., B. T. Grenfell, and T. H. Clutton-Brock. (2000) "Inverse density dependence and the Allee effect." *Trends in Ecology and Evolution* 14: 405–410. For reports on Iranian cheetah, see Iranian Cheetah Society, http://www.wildlife.ir/en /category/cheetah-monitoring.

6. For the role of biodiversity in ecosystem processes such as nutrient flow, see Schulze, E. D., and H. A. Mooney (eds.). (1993) *Biodiversity and Ecosystem Function*. Berlin: Springer-Verlag; Kinzig, A. P., S. W. Pacala, and D. Tilman (eds.). (2001) *The Functional Consequences of Biodiversity*. Princeton: Princeton University Press; Loreau, M., S. Naeem, and P. Inchausti (eds.). (2002) *Biodiversity and Ecosystem Functioning*. Oxford: Oxford University Press.

7. For the snow geese (*Chen caerulescens*) grazing story, see Jefferies, R. L., R. F. Rockwell, and K. F. Abraham. (2004) "Agricultural food subsidies, migratory connectivity and large-scale disturbance in Arctic coastal systems: a case study." *Integrative & Comparative Biology* 44: 130–139; Jefferies, R. L., A. P. Jano, and K. F. Abraham. (2006) "A biotic agent promotes large-scale catastrophic change in the coastal marshes of Hudson Bay." *Journal of Ecology* 94: 234–242.

8. The red-cockaded woodpecker (*Picoides borealis*) lives in mature pine forests of the southeastern United States. See James, F. C., C. A. Hess, and D. Kufrin. (1997) "Species-centered environmental analysis: indirect effects of fire history on red-cockaded woodpeckers." *Ecological Applications* 7: 118–129.

9. The orange-bellied parrot (*Neophema chrysogaster*) lives in southeast Australia and migrates to Tasmania to breed. See Stojanovic, D., F. Alves, H. Cook, R. Crates, R. Heinsohn, et al. (2018) "Further knowledge and urgent action required to save Orange-bellied Parrots from extinction." *Emu—Austral Ornithology* 118: 126–134. doi:10.1080/01584197.2017.1394165; see also Crates, R., L. Rayner, D. Stojanovic, M. Webb, and R. Heinsohn. (2017) "Undetected Allee effects in Australia's threatened birds: implications for conservation." *Emu—Austral Ornithology* 117: 207–221. doi:10.1080/01584197.2017.1333392.

10. For the consequences of man-eating lions, see Patterson, J. H. (1907) (repr. 1996). *The Maneaters of Tsavo*. New York: Pocket Books. See also Ford, J. (1971) *The Role of Trypanosomiases in African Ecology*. Oxford: Clarendon Press.

11. Simon Mduma recounts the problems he had hunting down man-eating lions in southern Tanzania in the 1970s. See pages 175–182 in Sinclair, A. R. E. (2012) *Serengeti Story*. Oxford: Oxford University Press.

12. Terborgh, J., and J. A. Estes (eds.). (2010) *Trophic Cascades: Predators, Prey, and the Changing Dynamics of Nature*. Washington, DC: Island Press. Also see note 1 in this chapter.

13. Coté, S. D., P. Rooney, J. P. Tremblay, C. Dussault, and D. K. Waller. (2004) "Ecological impacts of deer overabundance." *Annual Review of Ecology, Evolution, and Systematics* 35: 113–147; Ripple, W. J., T. P. Rooney, and R. L. Beschta. (2010) "Large predators, deer, and trophic cascades in boreal and temperate ecosystems." In Terborgh, J., and J. A. Estes (eds.). (2010) *Trophic Cascades: Predators, Prey, and the Changing Dynamics of Nature*. Washington, DC: Island Press: 141–162; Beschta, R. L., and W. J. Ripple. (2010) "Recovering riparian plant communities with wolves in Northern Yellowstone, U.S.A." *Restoration Ecology* 18: 380–389; Ripple, W. J., J. A.

Estes, R. L. Beschta, et al. (2014) "Status and ecological effects of the world's largest carnivores." *Science* 343: 6167. doi:10.1126/science. 1241484.

14. Angelstam, P., M. Manton, S. Pedersen, and M. Elbakidze. (2017) "Disrupted trophic interactions affect recruitment of boreal deciduous and coniferous trees in northern Europe." *Ecological Applications* 27: 1108–1123; Kuijper, D.P.J., C. de Kleine, M. Churski, P. van Hooft, J. Bubnicki, and B. Jedrzejewska. (2013) "Landscape of fear in Europe: wolves affect spatial patterns of ungulate browsing in Bialowieza Primeval Forest, Poland." *Ecography* 36: 1263–1275; Okarma, H. (1995) "The trophic ecology of wolves and their predatory role in ungulate communities of forest ecosystems in Europe." *Acta Theriologica* 40: 335–386.

15. For the influence of humans on the ecological role of predators, see Kuijper, D. P. J., E. Sahlen, B. Elmhagen, S. Chamaille-Jammes, H. Sand, K. Lone, and J. P. G. M. Cromsigt. (2016) "Paws without claws? Ecological effects of large carnivores in anthropogenic landscapes." *Proceedings of the Royal Society B* 283: article number 20161625. doi:10.1098/rspb.2016.1625.

16. For effects of removal of top predators in the ocean, see Baum, J. K, and B. Worm. (2009) "Cascading top-down effects of changing oceanic predator abundances." *Journal of Animal Ecology* 70: 699–714; Myers, R. A., and B. Worm. (2003) "Rapid worldwide depletion of predatory fish communities." *Nature* 423: 280–283; Myers, R. A., J. K. Baum, T. D. Shepherd, S. P. Powers, and C. H. Peterson. (2007) "Cascading effects of the loss of apex predatory sharks from a coastal ocean." *Science* 315: 1846–1850.

17. Jackson, J. B. C., M. X. Kirby, W. H. Berger, et al. (2001) "Historical overfishing and the recent collapse of coastal ecosystems." *Science* 293: 629–637. See also Sandin, S. A., S. M. Walsh, and J. B. C. Jackson. (2010) "Prey release, trophic cascades, and phase shifts in tropical nearshore ecosystems." In Terborgh, J., and J. A. Estes (eds.). (2010) *Trophic Cascades: Predators, Prey, and the Changing Dynamics of Nature.* Washington, DC: Island Press: 71–90.

18. The purple marsh crab is *Sesarma reticulatum.* See Silliman, B. R., and M. D. Bertness. (2002) "A trophic cascade regulates salt marsh primary production." *Proceedings of the National Academy of Sciences,USA* 99: 10500–10505; Altieri, A. H., M. D. Bertness, T. C. Coverdale, N. C. Herrmann, and C. Angelini. (2012) "A trophic cascade triggers collapse of a salt marsh ecosystem with intensive recreational fishing." *Ecology* 93: 1402–1410; Bertness, M. D., C. P. Brisson, T. C. Coverdale, M. C. Bevil, S. M. Crotty, and E. R. Suglia. (2014) "Experimental predator removal causes rapid salt marsh die-off." *Ecology Letters* 17: 830–835. doi:10.1111/ele.12287.

19. For Yellowstone National Park, see White, P. J, and R. A. Garrott. (2013) "Predation: Wolf restoration and the transition of Yellowstone elk." In White, P. J., R. A. Garrott, and G. E. Plumb (eds.), *Yellowstone's Wildlife in Transition.* Cambridge, MA: Harvard University Press: 69–93.

20. Brashares, J. S., L. R. Prugh, C. J. Stoner, and C. W. Epps. (2010) "Ecological and conservation implications of mesopredator release." In Terborgh, J., and J. A. Estes (eds.). (2010) *Trophic Cascades: Predators, Prey, and the Changing Dynamics of Nature.* Washington, DC: Island Press: 221–240.

21. See Terborgh, J. (1989) *Where Have All the Birds Gone?* Princeton: Princeton University Press; Crooks, K. R., and M. E. Soule. (1999) "Mesopredator release and avifaunal extinctions in a fragmented system." *Nature* 400: 563–566; Rogers, C. M., and M. J. Caro. (1998) "Song sparrows, top carnivores and nest predation: a test of the mesopredator release hypothesis." *Oecologia* 116: 227–233.

22. Terborgh, J., L. Lopez, P. Nunez, et al. (2001) "Ecological meltdown in predator-free forest fragments." *Science* 294: 1923–1926; Power, M. E., W. J. Matthews, and A. J. Stewart. (1985) "Grazing minnows, piscivorous bass, and stream algae: Dynamics of a strong interaction." *Ecology* 66: 1448–1456.

23. For killer whales, see Estes, J. A., M. T. Tinker, T. M. Williams, and D. E. Doak. (1998) "Killer whale predation on sea otters linking oceanic and nearshore ecosystems." *Science* 282: 473–476. For sea lions, see Trites, A. W., and C. P. Donnelly. (2003) "The decline of Steller sea lions *Eumetopias jubatus* in Alaska: a review of the nutritional stress hypothesis." *Mammal Review* 33: 3–28.

24. Rogers-Bennett, L., and C. A. Catton. (2019) "Marine heat wave and multiple stressors tip Bull Kelp forest to Sea Urchin barrens." *Scientific Reports* 9 (1): 15050.

25. For the island fox (*Urocyon littoralis*) and skunk (*Spilogale gracilis amphiala*) on Channel Islands, see Roemer, G. W., C. J. Donlan, and F. Courchamp. (2002) "Golden eagles, feral pigs, and insular carnivores: How exotic species turn native predators into prey." *Proceedings of the National Academy of Sciences, USA* 99: 791–796.

26. The brown-headed cowbird is *Molothrus ater,* the Least Bell's vireo is *Vireo bellii,* and the black-capped vireo is *V. atricapillus.* See Rothstein, S. I. (1994) "The cowbird's invasion of the far west: history, causes and consequences experienced by host species." *Studies in Avian Biology* 15: 201–315; Smith, J.N.M., T. L. Cook, S. I. Rothstein, S. K. Robinson, and S. G. Sealy (eds.). (2000) *Ecology and Management of Cowbirds and Their Hosts.* Austin: University of Texas Press.

27. For the effect of the Pacific rat (kiore) (*Rattus exulans*) on New Zealand ecosystems, see Atkinson, I.A.E., and E. K. Cameron. (1993) "Human influence on the terrestrial biota and biotic communities of New Zealand." *Trends in Ecology and Evolution* 8: 447–451; Cooper, A., I.A.E. Atkinson, W. G. Lee, and T. H. Worthy. (1993) "Evolution of the Moa and their effect on the New Zealand flora." *Trends in Ecology and Evolution* 8: 433–442; Campbell, D. J., and I. A. E. Atkinson. (1999) "Effects of kiore (*Rattus exulans* Peale) on recruitment of indigenous coastal trees on northern offshore islands of New Zealand." *Journal of the Royal Society of New Zealand* 29: 265–290; Hawke, D. J., R. N. Holdaway, J. E. Causer, and S. Ogden. (1999) "Soil indicators of pre-European seabird breeding in New Zealand at sites identified by predator deposits. *Australian Journal of Soil Research* 37: 103–113; Worthy, T. H., and R. N. Holdaway. (2002) *The Lost World of the Moa.* Bloomington: Indiana University Press.

28. Stan Boutin comments (November 2018) that the mountain caribou of British Columbia are affected by forest harvesting in the valleys, leading to more moose, more wolves, and higher predation rates on caribou. In Alberta, the increase in white-tailed deer due to warming temperatures from climate change has resulted in more wolves and thus increased predation. See Serrouya, R., D. R. Seip, D. Hervieux, et al. (2019) "Saving endangered species using adaptive management." *Proceedings of the National Academy of Sciences, USA* 116: 6181–6186. Also Wittmer, H. U., A. R. E. Sinclair, and B. N. McLellan. (2005) "The role of predation in the decline and extirpation of woodland caribou." *Oecologia* 144: 257–267.

29. Robley, A., J. Short, and S. Bradley. (2001) "Dietary overlap between the burrowing bettong (*Bettongia lesurur*) and the European rabbit (*Oryctolagus cuniculus*) in semi-arid coastal Western Australia." *Wildlife Research* 28: 341–349.

30. For extinction of marsupials, see Woinarski, J. C. Z., S. Legge, J. A. Fitzsimons, et al. (2011) "The disappearing mammal fauna of northern Australia: context, cause, and response."

Conservation Letters 4: 192–201; Fisher, D. O., C. N. Johnson, M. J. Lawes, et al. (2014) "The current decline of tropical marsupials in Australia: is history repeating?" *Global Ecology and Biogeography* 23: 181–190; Short, J., S. D. Bradshaw, J. Giles, R. I. T. Prince, and G. R. Wilson. 1992. Reintroduction of macropods (Marsupialia: Macropodoidea) in Australia: a review. *Biological Conservation* 62:189–204.

31. Reported to me by Norman Owen-Smith, 2003.

32. Reported to me by Richard Fynn, 2009.

33. For declines in Kruger Park, see Harrington, R. N., N. Owen-Smith, P. Viljoen, H. Biggs, and D. Mason. (1999) "Establishing the causes of the roan antelope decline in the Kruger National Park, South Africa." *Biological Conservation* 90: 69–78; du Toit, J. T., K. H. Rogers, and H. C. Biggs (eds.). (2003) *The Kruger Experience.* Washington, DC: Island Press. The collapse of the Botswana wildebeest migration is reported in Williamson, D., and B. Mbano. (1988) "Wildebeest mortality during 1983 at Lake Xau, Botswana." *African Journal of Ecology* 26: 341–344.

34. The collapse of the human-cattle migrations in the Sahel is described in Sinclair, A. R. E., and J. M. Fryxell. (1985) "The Sahel of Africa: ecology of a disaster." *Canadian Journal of Zoology* 63: 987–994.

35. When migrations are prevented, there is a rapid decline in population size. Migrations are discussed in Thirgood, S., A. Mosser, M. Borner, et al. (2004) "Can parks protect migratory ungulates? The case of the Serengeti wildebeest." *Animal Conservation* 7: 113–120. A review of migrations around the world appears in Harris, G., S. Thirgood, J. G. C. Hopcraft, J. P. G. M. Cromsigt, and J. Berger. (2009) "Global decline in aggregated migrations of large terrestrial mammals." *Endangered Species Research* 7: 55–76; Bolger, D. T., W. D. Newmark, T. A. Morrison, and D. F. Doak. (2008) "The need for integrative approaches to understand and conserve migratory ungulates." *Ecology Letters* 11: 63–77; Milner-Gulland, E. J., J. M. Fryxell, and A. R. E. Sinclair (eds.). (2011) *Animal Migration. A Synthesis.* Oxford: Oxford University Press.

36. For the noisy miner and eucalypt dieback, see Clarke, M. F., M. J. Grey, D. R. Britton, and R. H. Loyn. (1995) *The Noisy Miner,* Manorina melanocephala, *and rural dieback in remnant eucalypt woodlands.* Hawthorn East, Victoria, Australia: Royal Australian Ornithologists Union, No. 98. See also Landsberg, J. (1988) "Dieback of rural eucalypts: Tree phenology and damage caused by leaf-feeding insects." *Australian Journal of Ecology* 13: 251–267.

37. Rowan, R., N. Knowlton, A. Baker, and J. Jara. (1997) "Landscape ecology of algal symbionts creates variation in episodes of coral bleaching." *Nature* 388: 265–269.

38. For wildebeest effects on birds, see Nkwabi, A. K., A. R. E. Sinclair, K. L. Metzger, and S. A. R. Mduma. (2011) "Disturbance, ecosystem function and compensation: the influence of wildfire and grazing on the avian community in the Serengeti ecosystem, Tanzania." *Austral Ecology* 36: 403–412; Nkwabi, A. K., A. R. E. Sinclair, and S. A. R. Mduma. (2015). "The effect of natural disturbances on the avian community of the Serengeti woodlands." In A. R. E. Sinclair, K. L. Metzger, J. M. Fryxell, and S. A. R. Mduma (eds.), *Serengeti IV: Sustaining Biodiversity in a Coupled Human-Natural System.* Chicago: University of Chicago Press: 395–418.

39. Modeling of communities suggests that the greater the degree of interactions between species, the greater the stability of the system. It also suggests that the more species are lost, the greater the instability becomes, but this applies most to keystone species and least to insignificant species. See de Visser, S. N., B. P. Freymann, and H. Olff. (2011) "The Serengeti food web:

Empirical quantification and analysis of topological changes under increasing human impact." *Journal of Animal Ecology* 80: 484–494; Dunne, J. A., R. J. Williams, and N. D. Martinez. (2002) "Network structure and biodiversity loss in food webs: robustness increases with connectance." *Ecology Letters* 5: 558–567.

40. Sutherland, W. J. (1998) "Managing habitats and species." In W. J. Sutherland (ed.). *Conservation Science and Action.* Oxford: Blackwell Science: 202–219.

41. For fire effects, see Williams, J. E., R. J. Whelan, and A. M. Gill. (1994) "Fire and environmental heterogeneity in southern temperate forest ecosystems: implications for management." *Australian Journal of Botany* 42: 125–137.

42. Bush, M. B., and P. A. Colinvaux. (1994) "Tropical forest disturbance: paleoecological records from Darien, Panama." *Ecology* 75: 761–768.

43. Bergstrom, D. M., P. K. Bricher, B. Raymond, et al. (2015) "Rapid collapse of a sub-Antarctic alpine ecosystem: the role of climate and pathogens." *Journal of Applied Ecology* 52: 774–783.

44. Behling, H., and V. DePatta Pillar. (2007) "Late Quaternary vegetation, biodiversity and fire dynamics on the southern Brazilian highland and their implication for conservation and management of modern *Araucaria* forest and grassland ecosystems." *Philosophical Transactions of the Royal Society, B.* 362: 243–251.

45. For the giant pandas (*Ailuropoda melanoleuca*) in China, see Schaller, G., H. Jinchu, P. Wenshi, and Z. Zing. (1985) *The Giant Pandas of Wolong.* Chicago: University of Chicago Press; Schaller, G. B., T. Qitao, K. G. Johnson, W. Xiaoming, S. Heming, and H. Jinchu. (1989) "The feeding ecology of Giant Pandas and Asiatic Black Bears in the Tangjiahe Reserve, China." In J. L. Gittleman (ed.), *Carnivore Behaviour, Ecology, and Evolution.* London: Chapman and Hall: 212–241.

46. Woodroffe, R. (1998) "Edge Effects and the Extinction of Populations inside Protected Areas." *Science* 280: 2126–2128.

47. Effects of global climate change on ecosystems can be found in McCarty, J. P., L. L. Wolfenbarger, and J. A. Wilson. (2017). "Biological Impacts of Climate Change." In *ELS (Encyclopedia of Life Sciences)*, Oxford: John Wiley & Sons, Ltd. (Ed.). doi:10.1002/9780470015902. a0020480.pub2; Pounds, J. A., M. P. L. Fogden, and J. H. Campbell. (1999) "Biological response to climate change on a tropical mountain." *Nature* 398: 611–615; Schneider, S. H., and T. L. Root (eds.). (2002) *Wildlife Responses to Climate Change.* Washington, DC: Island Press; Parmesan, C., and G. Yohe. (2003) "A globally coherent fingerprint of climate change impacts across natural systems." *Nature* 421: 37–42. Northward movements in species ranges are documented in Thomas, C. D., and J. J. Lennon. (1999) "Birds extend their ranges northwards." *Nature* 399: 213. For changes in species elevation range, see Wilson, R. J., and D. Gutierrez. (2012) "Effect of climate change on the elevational limits of species ranges." In E. A. Beever, and J. L. Belant (eds.), *Ecological Consequences of Climate Change.* Boca Raton, FL: CRC Press: 107–132. Also Moritz, C., J. L. Patton, C. Conroy, J. L. Parra, G. C. White, and S. R. Beissinger. (2008) "Impact of a century of climate change on small-mammal communities in Yosemite National Park, USA. *Science* 322: 261–264.

48. Examples of the change in timing of breeding in birds are Visser, M., and C. Both. (2005) "Shifts in phenology due to global climate change: the need for a yardstick." *Proceedings of the Royal Society of London, B* 272: 2561–2569; Visser, M., J. M. Holleman, and P. Gienapp. (2006)

"Shifts in caterpillar biomass phenology due to climate change and its impact on the breeding biology of an insectivorous bird." *Oecologia* 147: 164–172; Reed, T. E., S. Jenouvrier, and M. E. Visser. (2013) "Phenological mismatch strongly affects individual fitness but not population demography in a woodland passerine." *Journal of Animal Ecology* 82: 131–144. For timing of arrival of migrant birds, see Beaumont, L. J., I. A. W. McAllan, and L. Hughes. (2006) "A matter of timing: Changes in the first date of arrival and last date of departure of Australian migratory birds." *Global Change Biology* 12: 1339–1354; Root, T. L., J. T. Price, K. R. Hall, S. H. Schneider, C. Rosenzweig, and J. A. Pounds. (2003) "Fingerprints of wild animals and plants." *Nature* 421: 57–60. See also Drever, M. C., R. G. Clark, C. Derksen, S. M. Slatterly, P. Toose, and T. D. Nudds. (2012) "Population vulnerability to climate change linked to timing of breeding of boreal ducks." *Global Change Biology* 18: 480–492.

49. The account of soil salinization can be found in Hobbs, R. J. (1993) "Effects of landscape fragmentation on ecosystem processes in the Western Australia wheatbelt." *Biological Conservation* 64: 193–201; McFarlane, D. J., R. J. George, and P. Farrington. (1993) "Changes in the Hydrologic Cycle." In R. J. Hobbs and D. A. Saunders (eds.), *Reintegrating Fragmented Landscapes.* New York: Springer-Verlag: 147–186.

50. For high-Arctic disturbance, see Becker, M. S., and W. H. Pollard. (2016) "Sixty-year legacy of human impacts on a high Arctic ecosystem." *Journal of Applied Ecology* 53: 876–884.

51. See note 34. For the Sahara cycle, see Otterman, J. (1974) "Baring high-albedo soils by overgrazing: a hypothesized desertification mechanism." *Science* 186: 531–533.

52. For Niger, see Wu, X. B., T. L. Thurow, and S. G. Whisenant. (2000) "Fragmentation and changes in hydrologic function of tiger bush landscapes, south-west Niger." *Journal of Animal Ecology* 88: 790–800.

53. For Brandt's vole in China, see Zhang, Z., R. Pech, S. Davis, D. Shi, X. Wan, and W. Zhong. (2003) "Extrinsic and intrinsic factors determine the eruptive dynamics of Brandt's voles *Microtus brandti* in Inner Mongolia, China." *Oikos* 100: 299–310.

54. Jiang, J., S. Ni, A. Gong, W. Wang, Y. Zha, J. Wang, and F. Voss. (1999) "Grassland degradation and its control in the region around Qinghai Lake." *Pedosphere* 9: 331–338.

55. See note 53.

56. For keystones determining biodiversity, see Paine, R. T. (1966) "Food web complexity and species diversity." *American Naturalist* 100: 65–75; Paine, R. T. (1980) "Food webs: linkage, interaction strength and community infrastructure." *Journal of Animal Ecology* 49: 667–685; Connell, J. H. (1978) "Diversity in tropical rainforest and coral reefs." *Science* 199: 1302–1310.

57. Walker, B. H., D. Ludwig, C. S. Holling, and R. M. Peterman. (1981) "Stability of semi-arid savanna grazing systems." *Journal of Ecology* 69: 473–498.

58. For the role of invasive species, see Vitousek, P. M., C. M. D'Antonio, L. L. Loope, and R. Westbrooks. (1996) "Biological invasions as global environmental change." *American Scientist* 84: 468–478.

59. Reported by Bill Robichaud (director of the Saola Foundation) and Grant Singleton in Sinclair, A. R. E., and A. Byrom. (2006) "Understanding ecosystems for the conservation of biota." *Journal of Animal Ecology* 75: 64–79.

60. The Atlantic cod is *Gadus morhua*. Reported by Ray Hilborn, October 2018. See also Choi, J. S., K. T. Frank, W. C. Leggett, and K. Drinkwater. (2004) "Transition to an alternate state in a continental shelf system." *Canadian Journal of Fisheries and Aquatic Science* 61: 505–510;

Frank, K. T., B. Petrie, and W. C. Leggett. (2005) "Trophic cascades in a formerly cod-dominated ecosystem." *Science* 308: 1621–1623.

61. Hughes, T. P. (1994) "Catastrophes, phase shifts, and large-scale degradation of a Caribbean coral reef." *Science* 265: 1547–1551.

62. A useful summary of the views concerning the die-off of the megaherbivores at the end of the Pleistocene is given in MacPhee, R. D. E. (2019) *The End of the Megafauna*. New York: W. W. Norton & Co.

63. For a description of the events surrounding the Fortymile caribou herd, see chapter 7, page 133 in Fryxell, J. M., A. R. E. Sinclair and G. Caughley. (2014) *Wildlife Ecology, Conservation, and Management*. Oxford: Wiley Blackwell, 3rd ed. See also Boertje, R. D., C. L. Gardner, M. M. Ellis, T. W. Bentzen, and J. A. Gross. (2017) "Demography of an increasing caribou herd with restricted wolf control." *Journal of Wildlife Management* 81: 429–448.

64. The collapse of the Saiga antelope in Asia was reported in Milner-Gulland, E. J., M. V. Kholodova, A. Bekenov, O. M. Bukreeva, I. A. Grachev, L. Amgalan, and A. A. Lushchekina. (2001) "Dramatic Declines in Saiga Antelope Populations." *Oryx* 35: 340–345. The collapse of other species and the increase in wolves is reported in Bragina, E. V., A. R. Ives, A. M. Pidgeon, et al. (2015) "Rapid declines of large mammal populations after the collapse of the Soviet Union." *Conservation Biology* 29: 844–853.

65. For a discussion of greater variability of weather due to climate change, see Kareiva, P. M. J., J. G. Kingsolver, and R. B. Huey (eds.). (1993) *Biotic Interactions and Global Change*. Sunderland, MA: Sinauer; Hellmann, J. J. (2002) "Butterflies as model systems for understanding and predicting climate change." In S. H. Schneider and T. L. Root (eds.), *Wildlife Responses to Climate Change*. Washington, DC: Island Press: 93–126.

66. Broecker, W. S. (1987) "Unpleasant surprises in the greenhouse?" *Nature* 328: 123–126; de la Mare, W. K. (1997) "Abrupt mid-twentieth-century decline in Antarctic sea-ice extent from whaling records." *Nature* 389: 57–60; Atkinson, A., V. Siegel, E. Pakhomov, et al. (2004) "Long-term decline in krill stock and increase in salps within the Southern Ocean." *Nature* 432: 100–103. https://doi.org/10.1038/nature02996.

67. Information provided by Dr. M. Julian Caley, Townsville, Queensland. See MacNeil, M. A., C. Mellin, S. Matthews, et al. (2019) "Water quality mediates resilience on the Great Barrier Reef." *Nature Ecology & Evolution* 3: 620–627.

68. For research in Gorongosa National Park, Mozambique, see Guyton, J. A., J. Pansu, M. C. Hutchinson, et al. (2020) "Trophic rewilding revives biotic resistance to shrub invasion." *Nature Ecology & Evolution* 4: 712–724; Branco, P. S., J. A. Merkle, R. M. Pringle, et al. (2019) "Determinants of elephant foraging behavior in a coupled human-natural system: is brown the new green?" *Journal of Animal Ecology* 88: 780–792; Stalmans, M. E., T. J. Massad, M. J. S. Peel, C. E. Tarnita, and R. M. Pringle. (2019) "War-induced collapse and asymmetric recovery of large-mammal populations in Gorongosa National Park, Mozambique." *PLoS ONE* 14: e0212864; Atkins, J. L., R. A. Long, J. Pansu, et al. (2019) "Cascading impacts of large-carnivore extirpation in an African ecosystem." *Science* 364: 173–177; Branco, P. S., J. A. Merkle, R. M. Pringle, L. King, T. Tindall, M. Stalmans, and R. A. Long. (2019) "An experimental test of community-based strategies for mitigating human-wildlife conflict around protected areas." *Conservation Letters* 2019: e12679.

69. Bhattachan, A., P. D'odorico, K. Dintwe, G. S. Okin, and S. L. Collins. (2014) "Resilience and recovery potential of duneland vegetation in the southern Kalahari." *Ecosphere* 5: 1–14.

70. For Banff National Park, Canada, see Hebblewhite, M., C. A. White, C. G. Clifford, G. Nietvelt, J. A. Mckenzie, T. E. Hurd, J. M. Fryxell, S. E. Bayley, and P. C. Paquet. (2005) "Human activity mediates a trophic cascade caused by wolves." *Ecology* 86: 2135–2144.

Chapter 14: Rewilding

1. Estes, J. A., J. Terborgh, J. S. Brashares, et al. (2011) "Trophic Downgrading of Planet Earth." *Science* 333 (6040): 301.

2. World Wildlife Fund. (2018) "Living Planet Report—2018: Aiming Higher." Grooten, M., and R. E. A. Almond (eds.). Gland, Switzerland: WWF.

3. Ceballos, G., P. R. Ehrlich, A. D. Barnosky, A. Garcia, R. M. Pringle, and T. M. Palmer. (2015) "Accelerated modern human-induced species losses: Entering the sixth mass extinction." *Science Advances* 1 (5): e1400253.

4. Ceballos, G., and P. R. Ehrlich. (2018) "The misunderstood sixth mass extinction." *Science* 360: 1080–1081; Davis, M., S. Faurby, and J.-C. Svenning. (2018) "Mammal Diversity Will Take Millions of Years to Recover from the Current Biodiversity Crisis." *Proceedings of the National Academy of Sciences, USA* 155 (44):11262–11267.

5. Ceballos, G. (2017) "Biological Annihilation via the Ongoing Sixth Mass Extinction Signaled." *Proceedings of the National Academy of Science, USA* 114 (30): E6089–E6096.

6. The Intergovernmental Science-Policy Platform on Biodiversity and Ecosystem Services, in its Global Assessment Report on Biodiversity and Ecosystem Services (May 2019), estimates that a million species of animals and plants may be threatened with extinction. The report can be found at https://www.ipbes.net/news/Media-Release-Global-Assessment.

7. See Brashares, J. S., P. Arcese, M. K. Sam, P. B. Coppolillo, A. R. E. Sinclair, and A. Balmford. (2004) "Bushmeat Hunting, Wildlife Declines, and Fish Supply in West Africa." *Science* 306 (5699):1180–1183; Western, D., S. Russell, and I. Cuthill. (2009) "The Status of Wildlife in Protected Areas Compared to Non-Protected Areas of Kenya." *PLoS ONE* 4 (7): e6140.

8. Hallmann, C. A., M. Sorg, E. Jongejans, et al. (2017) "More than 75 percent decline over 27 years in total flying insect biomass in protected areas." *PLoS ONE* 12 (10): e0185809.

9. See https://www.protectedplanet.net.

10. Council on Environmental Quality. (1974) "Environmental Quality." Washington, DC: Council on Environmental Quality, 5th Annual Report; Wade, N. (1974) "Sahelian drought: no victory for western aid." *Science* 185: 234–237.

11. For the Carrifran rewilding project, see Ashmole, P., and M. Ashmole (eds.). (2020) *A Journey in Landscape Restoration: Carrifran Wildwood and Beyond.* Caithness, Scotland: Whittles Publishing; http://www.carrifran.org.uk; Savory, C. J. (2016) "Colonisation by woodland birds at Carrifran Wildwood: the story so far." *Scottish Birds* 36: 135–149. For examples of rewilding in Wales, see https://www.cambrianwildwood.org.

12. Hutton, I., J. P. Parkes, and A. R. E. Sinclair. (2007) "Re-assembling island ecosystems: the case of Lord Howe Island." *Animal Conservation* 10: 22–29.

13. Noble, J. C., D. S. Hik, and A. R. E. Sinclair. (2007) "Landscape ecology of the burrowing bettong: fire and marsupial biocontrol of shrubs in semi-arid Australia." *Rangelands Journal* 29: 107–119.

14. On 11 June 1996, Aboriginal trackers showed me and colleagues Roger Pech and Alan Newsome the likely habitat, based on traditional knowledge, for release of rufous hare-wallaby (*Lagorchestes hirsutus*), or mala, in the Tanami Desert of the Northern Territory of Australia, which eventually took place in 2019.

15. See their website: https://treesforlife.org.uk. See also Ashmole, P., and M. Ashmole (eds.). (2020) *A Journey in Landscape Restoration: Carrifran Wildwood and Beyond.* Caithness, Scotland: Whittles Publishing, http://www.carrifran.org.uk.

16. For information about reintroduction of lynx in Scotland, see https://scottishwildlifetrust .org.uk/2019/07/could-we-reintroduce-lynx-to-Scotland.

17. Beschta, R. L., and W. J. Ripple. (2019) "Can large carnivores change streams via a trophic cascade?" *Ecohydrology* 12 (1): UNSP e2048; Ripple, W. J., and R. L. Beschta. (2012) "Trophic Cascades in Yellowstone: The First 15 years after Wolf Reintroduction." *Biological Conservation* 145 (1): 205–213; Creel, S., J. A. Winnie, B. Maxwell, K. Hamlin, and M. Creel. (2005) "Elk alter habitat selection as an antipredator response to wolves." *Ecology* 86: 3387–3397.

18. For estimates of the numbers of golden jackal, wolf, and other large carnivores in Europe, see https://www.lcie.org/Largecarnivores.

19. Chapron, G., P. Kaczensky, J. D. C. Linnell, et al. (2014) "Recovery of large carnivores in Europe's modern human-dominated landscapes." *Science* 346: 1517–1519.

20. Zamboni, T., S. Di Martino, and I. Jimenez-Perez. (2017) "A review of a multispecies reintroduction to restore a large ecosystem: The Iberá Rewilding Program (Argentina)." *Perspectives in Ecology and Conservation* 15 (4): 248–256.

21. Pringle, R. M. (2017) "Upgrading protected areas to conserve wild biodiversity." *Nature* 546: 91–99.

22. https://y2y.net.

23. Thapa, K., E. Wikramanayake, S. Malla, et al. (2017) "Tigers in the Terai: Strong evidence for meta-population dynamics contributing to Tiger recovery and conservation in the Terai Arc Landscape." *PLoS ONE*, 12 (6): e0177548.

24. https://www.sanparks.org/about/news/?id=56251.

25. This connectivity project is presented as one of 25 case studies around the world in Hilty, J., G. L. Worboys, A. Keeley, et al. (2020) "Guidelines for conserving connectivity through ecological networks and corridors." Gland, Switzerland: Best Practice Protected Area Guidelines Series No. 30. IUCN.

26. Fraser, C. (2009) *Rewilding the World: Dispatches from the Conservation Revolution.* New York: Henry Holt and Co.

27. See https://www.panthera.org/initiative/jaguar-corridor-initiative and Hilty, J., et al. in note 25 above.

28. Gaywood, M. J. (2017) "Reintroducing the Eurasian Beaver, *Castor fiber*, to Scotland." *Mammal Review* 48: 48–61. doi:10.1111/mam.12113.

29. Pollock, M. M., G. Lewallen, K. Woodruff, C. E. Jordan, and J. M. Castro (eds.). (2015) *The Beaver Restoration Guidebook: Working with Beaver to Restore Streams, Wetlands, and Floodplains. Version 1.02.* Portland, OR: United States Fish and Wildlife Service. https://www.fws.gov /oregonfwo/promo.cfm?id=177175812.

30. Hansen, D. M. (2015) "Non-native megaherbivores: the case for novel function to manage plant invasions on islands." *Annals of Botany Plants* 7 (SI): plv085; Hansen, D. M.,

C. J. Donlan, C. J. Griffiths, and K. J. Campbell. (2010) "Ecological history and latent conservation potential: large and giant tortoises as a model for taxon substitutions." *Ecography* 33: 272–284.

31. According to the theory of island biogeography, the number of species on an island depends on the size of the area and the distance to the mainland (or other islands) from which new species can immigrate. See MacArthur, R. H., and E. O. Wilson. (1967) *The Theory of Island Biogeography*. Princeton: Princeton University Press. Extinction debt is discussed in Tilman, D. (1994) "Habitat Destruction and the Extinction Debt." *Nature* 371 (6492): 65–66. Faunal relaxation was first mentioned in Diamond, J. M. (1972) "Biogeographic kinetics—estimation of relaxation times for avifaunas of Southwest Pacific islands." *Proceedings of the National Academy of Sciences, USA* 69: 3199–3203.

32. Brashares, J. S., P. Arcese, and M. K. Sam. 2001 "Human demography and reserve size predict wildlife extinction in West Africa." *Proceedings of the Royal Society London* 268: 2473–2478.

33. Natural Resources Canada: https://www.nrcan.gc.ca/our-natural-resources/forests-forestry/wildland-fires-insects-disturban/top-forest-insects-diseases-cana/mountain-pine-beetle/13381.

34. Boisramé, G. F. S., S. E. Thompson, C. N. Tague, and S. L. Stephens. (2019) "Restoring a natural fire regime alters the water balance of a Sierra Nevada catchment." *Water Resources Research* 55 (7): 5751–5769.

35. Kremen, C., and A. M. Merenlender. (2018) "Landscapes that work for biodiversity and people." *Science* 362 (6412): eaau6020.

36. Saunders, G. R., M. N. Gentle, and C. R. Dickman. (2010) "The impacts and management of foxes, *Vulpes vulpes*, in Australia." *Mammal Review* 40: 181–211.

37. Hone, J. (1999) "Fox control and rock-wallaby population dynamics—assumptions and hypotheses." *Wildlife Research* 26: 671–673; Kinnear, J. E., M. L. Onus, and R. N. Bromilow. (1988) "Fox control and rock-wallaby population dynamics." *Wildlife Research* 15: 435–450; Kinnear, J. E., M. L. Onus, and N. R. Sumner. (1998) "Fox control and rock-wallaby population dynamics—II. An update." *Wildlife Research* 25: 81–88; Kinnear, J. E., N. R. Sumner, and M. L. Onus. (2002) "The red fox in Australia—an exotic predator turned biocontrol agent." *Biological Conservation* 108: 335–359.

38. Sinclair, A. R. E., R. P. Pech, C. R. Dickman, D. Hik, P. Mahon, and A. E. Newsome. (1998) "Predicting effects of predation on conservation of endangered prey." *Conservation Biology* 12: 564–575; Binny, R. N., J. Innes, N. Fitzgerald, R. Pech, A. James, R. Price, C. Gillies, and A. E. Byrom. (2021) "Long-term biodiversity trajectories for pest-managed ecological restorations: eradication versus suppression." *Ecological Monographs* 91 (2): e01439. doi:10.1002/ecm.1439.

39. Ripple, W. J., E. J. Larsen, R. A. Renkin, and D. W. Smith. (2001) "Trophic cascades among wolves, elk and aspen on Yellowstone National Park's northern range." *Biological Conservation* 102: 227–234; Ripple, W. J., T. P. Rooney, and R. L. Beschta. (2010) "Large predators, deer and trophic cascades in boreal and temperate ecosystems." In Terborgh, J., and J. A. Estes (eds.), *Trophic Cascades: Predators, Prey, and the Changing Dynamics of Nature*. Washington, DC: Island Press: 141–162.

40. Hebblewhite, M., C. A. White, C. G. Clifford, et al. (2005) "Human activity mediates a trophic cascade caused by wolves." *Ecology* 86: 2135–2144.

41. White, P. J., R. A. Garrott, and G. E. Plumb (eds.). (2013) *Yellowstone's Wildlife in Transition*. Cambridge, MA: Harvard University Press.

42. Foote, J. (1990) "Trying to take back the planet." *Newsweek* 115 (6): 24.

43. Soulé, M., and R. Noss. (1998) "Rewilding and Biodiversity: Complementary Goals for Continental Conservation." *Wild Earth* 11: 18–28.

44. Foreman, D. (2004) *Rewilding North America: A Vision for Conservation in the 21st Century*. Washington, DC: Island Press.

45. Fraser, L. H., W. L. Harrower, H. W. Garris, et al. (2015) "A Call for Applying Trophic Structure in Ecological Restoration." *Restoration Ecology* 23: 503–507.

46. McDonald, T., G. D. Gann, J. Jonson, K. W. Dixon, and W. Kingsley. (2016) *International Standards for the Practice of Ecological Restoration—Including Principles and Key Concepts*. Society for Ecological Restoration (SER), 1st ed.

47. Jørgensen, D. (2014) "Rethinking rewilding." *Geoforum* 65: 482–488; Higgs, E., D. A. Falk, A. Guerrini, et al. (2014) "The changing role of history in restoration ecology." *Frontiers in Ecology and the Environment* 12 (9): 499–506.

48. Pettorelli, N., S. Durant, and J. du Toit (eds.). 2019 *Rewilding (Ecological Reviews)*. Cambridge: Cambridge University Press.

49. Choi, Y. D. (2007) "Restoration ecology to the future: A call for new paradigm." *Restoration Ecology* 15 (6): 351–353.

50. Stuart, A. J. (2021) *Vanished Giants: The Lost World of the Ice Age*. Chicago, University of Chicago Press.

51. Donlan, J., J. Berger, C. E. Bock, et al. (2005) "Re-wilding North America." *Nature* 436 (7053): 913–914.

52. Oliveira-Santos, L. R. G., and F. A. S. Fernandez. (2010). "Pleistocene rewilding, Frankenstein ecosystems, and an alternative conservation agenda." *Conservation Biology* 24: 4–6; Rubenstein, D. R., and D. I. Rubenstein. (2016) "From Pleistocene to Trophic Rewilding: A Wolf in Sheep's Clothing." *Proceedings of the National Academy of Sciences, USA* 113 (1): E1; Rubenstein, D. R., D. I. Rubenstein, P. W. Sherman, and T. A. Gavin. (2006) "Pleistocene park: Does re-wilding North America represent sound conservation for the 21st century?" *Biological Conservation* 132 (2): 232–238; Carey, J. (2016) "Core Concept: Rewilding." *Proceedings of the National Academy of Sciences, USA* 113 (4): 806–808.

53. Vera, F. W. (2009) "Large-scale nature development—The Oostvaardersplassen." *British Wildlife* 20: 28. Svenning, J.-C., P. B. M. Pedersen, C. J. Donlan, et al. (2016) "Science for a Wilder Anthropocene: Synthesis and Future Directions for Trophic Rewilding Research." *Proceedings of the National Academy of Sciences, USA* 113 (4): 898–906. See Pettorelli, N., et al. (2019) in note 48 above.

54. See note 12.

55. Soule, M. E., J. A. Estes, J. Berger, and C. Martinez del Rio. (2003) "Ecological Effectiveness: Conservation Goals for Interactive Species." *Conservation Biology* 17 (5): 1238–1250.

56. Sinclair, A. R. E., R. P. Pech, J. M. Fryxell, et al. (2018) "Predicting and Assessing Progress in the Restoration of Ecosystems." *Conservation Letters* 11 (2): UNSP e12390.

INDEX

aardvark, 20, 154

aardwolf, 154; *Trinervitermes* (termite) and, 99

Acacia species, 22; acacia savanna, 10–11, 224; acacia trees, 103

aerial photographs: sampling technique using, 35–36, 40; used to count wildebeest and buffalo, 33–34; view of buffalo herd, 35

aestivation, 111

African buffalo, 18

African Humid Period, 248n10

African Rift Valley, 116, 258n30

Afrotheria, 20, 226n15

agriculture, 159–60; and fertilizers in lakes and rivers, 174; and habitat modification, 169; and impact of humans on plant communities, 188

Alaska: coast of, 177; moose in, 179

Albertine Rift, 7, 8

Allee, W. C., 173, 260n4

Allee effect, 152, 173, 260n5

alternative stable states: as seventh subprinciple, 147, 170–71, 186–90; unwanted, 186–90

Amazon, deforestation of, 187

American University, 195, 197

Anderson, Mike, 148, 149, 150, 151, 154

Andrewartha, H. G., 37, 38

animals: and camera traps, 90; cause of death in, 44–45; death by predators of, 61–62; fat stores in living, 48; habitats for, 90–91; marrow fat of, 48–49; and regulation hypothesis, 38; and role of

predators, 48–49; use of aerial photographs to count, 25

animals in Serengeti, 13–21; African buffalo, 18; antelope species, 17–18; biodiversity of, 21; giraffe, 18; ground pangolin, 19; hyenas, 18–19; lions, 18; migrants, 16–18; residents, 16–18; taxonomy of using DNA comparisons, 20; Thomson's gazelle, 14–15; wild dogs, 19; wildebeest, 13–16; zebra, 13–16

antelopes: collapse of Saiga, 266n64; food supply and, 60; percentage killed by predators, 63; regulation of, 59–61; species of, 17–18; use of radio collars on, 61–62

Aparados da Serra National Park, 184

apparent competition, 241n6

Aral Sea, 158, 255n5

Araucaria forest, 184–85

Arcese, Peter, 61

army worm moth (*Spodoptera exigua*), 245n12

Artiodactyla, 221–22

Asian cheetah, 173

Atlantic cod (*Gadus morhua*), 189, 265n60

Australia: agricultural decline in, 186; burrowing bettong in, 179–80; competition and predation in, 180; Great Barrier Reef of, 190; honeyeaters in, 182; nonnative predators in, 208; rufous hare-wallaby (*Lagorchestes hirsutus*) in, 268n14

Azorella macquariensis (keystone cushion plant), 184

A NOTE ON THE TYPE

This book has been composed in Arno, an Old-style serif typeface in the classic Venetian tradition, designed by Robert Slimbach at Adobe.